Amazon Echo

The Complete User Guide:
Learn to Use Your Echo Like A Pro

by C.J. Andersen

Copyright © 2019 by CJ Andersen — All rights reserved.

AMAZON ECHO is a trademark of Amazon Technologies Inc. All other copyrights & trademarks are the properties of their respective owners. Reproduction of any part of this book without the permission of the copyright owners is illegal — the only exception to this is the inclusion of brief quotes in a review of the work. Any request for permission should be directed to cjandersentech@gmail.com

Contents

Introduction

Before we get Started: Echo Basics — 8
   The Echo Preview — 9
   Specifications of the Amazon Echo — 11

1: Let's Begin! —Echo Setup — 13
   Downloading the Alexa App — 13
   Pairing Mobile Devices with Echo Show — 15
   Creating Your Voice Profile — 17
   Creating an Amazon Household — 18

2: Meet the Alexa App — 20
   The App and your Voice — 20
   A Word About the Alexa App Layout — 21
   The Alexa App Homepage on your PC or Mac — 21
   The Alexa App Homepage on your mobile device — 24

3: The 'Now Playing' Homepage and Player — 26

4: Watch this and More! — Video — 30
   Fire TV — 31
   Dish TV — 32

5: Songs by the Millions — Music — 34
   My Music Library/ My Amazon Music — 35
   Amazon Prime Music — 38
   Add Prime Playlists to My Music Library — 40
   Spotify — 42
   Pandora — 43
   iHeartRadio — 45
   Tunein — 47
   SiriusXM — 49
   Deezer — 51
   Other Music/ Radio Skills — 51
   Setting Default Music Service — 52
   Setting Up Multi-Room Music — 53
   Setting Up Stereo Sound Speaker Sets — 54

6: Tell Me a Story — Books — 55
Audible — 55
Kindle Books — 57

7: Write it Down & Get it Done — Lists — 60
Shopping Lists — 60
To-do Lists — 62
Create Your Own List — 64
Delete a List — 64

8: Never Forget, Always Be On Time — Reminders, Alarms & Timers — 65
Reminders — 65
Alarms — 67
Timers — 69

9: Keep it Simple – Routines — 70

10: Let's See What She Can Do — Alexa Skills — 74
Get Familiar with Skills — 74
Enable Skills and Use Them — 77
Disable a Skill — 78
Alexa Skills Blueprint — 79

11: 21st Century Living — Smart Home — 81
Set Up Smart Home Devices — 82
Set Up Groups and Scenes for Your Smart Home Devices — 83
Troubleshooting Alexa and Smart Home Device Set Up — 85

12: Your Highly Capable Personal Assistant — Things to Try — 88
What's new? — 89
Ask Questions — 90
Check and Manage Your Calendar — 90
Discover New Alexa Skills — 90
Find Local Businesses and Restaurants & Reviews — 91
Find Traffic Information — 92
Get Weather Forecasts — 92
Go to the Movies — 93
Hear the News – Flash Briefing — 93
Get Your Sports Updates — 94

Shop Amazon	95
Calling and Messaging	97
Drop In	99
Announcements	100
Kids Skills & Safety	101
13: Alexa & Echo: Your Perfect Fit — Settings	**102**
Device Settings	102
Do Not Disturb	103
Sounds	103
Wake Word	104
Setting Alexa Preferences	105
Choose Default Music Services	105
Flash Briefing	106
Alexa Account Settings	108
Voice Purchasing	108
Household Profile	108
Managing and Deleting Voice Recordings	109
14: Resources Galore — Help & Feedback	**110**
15. Going to the Source – Amazon Pages to Be Familiar With	**113**
Before You Go	**114**

Introduction

Hello and welcome! Thank you for buying this Amazon Echo user guide and I look forward to sharing my knowledge and helping you to master your new Amazon device. Before we get under way I would like to make a couple of suggestions. Firstly, I would like to encourage you to get familiar with Amazon's Help Pages (https://amzn.to/2IObbJu) and here's why... Amazon is constantly tweaking, updating and expanding the capabilities of both its devices (Fire tablets, Fire TV, Echo, Echo Dot, Echo Show etc) and its AI software Alexa.

And when I say constantly, I mean ALL the time.

While I will continue to publish updated editions of this guide, I want to be sure that anyone who has bought this book has the chance to stay up to date with the frequent changes and Amazon's help pages for your device is the best place to do that. When changes to the way your device works or new features are introduced, the information is also updated on these help pages. Personally, I don't find their help pages super user-friendly, but nevertheless they are the best place to stay up to date with changes and improvements.

Secondly, I wanted to say a word about this guide, how it has been put together and the best way to use it. Of course, you should feel free to go straight to a particular chapter if there is some specific information you are looking for, but I would also urge you to read the book in its entirety as it follows a logical progression.

If this is your first Alexa-enabled device, then you will discover that the Alexa App is at the heart of everything you can do with the Echo. For that reason, I have laid out this guide to take you step by step through every option within the App starting at the top and working down to the bottom. It's really the easiest way to approach and master Alexa without jumping back and forth and getting lost in the numerous options.

I'll also be reviewing the many third-party apps and Alexa Skills that you can access via the Echo to maximize your enjoyment of the device, including:

- Using music apps such as Amazon Music, Spotify, Tunein and iHeartRadio
- Linking up to audiobook services like Audible and Kindle
- Controlling video streaming devices such as the Fire TV
- Setting up the Echo as an integral part of your Smart Home technology
- Getting the most out of the Alexa personal assistant, from setting alarms to ordering pizzas!
- And, of course, troubleshooting common glitches that crop up from time to time...

I am confident that anyone, no matter what level their tech skills, will find learning and using their Amazon Echo pretty straightforward. If, after reading this guide, there's anything that you still find confusing, then please email me at cjandersentech@gmail.com

Okay, let's go!

Before we get started: Echo Basics

I was an enthusiastic early-adopter of the original Amazon Echo, and have been very happy with its performance. And I'm not the only one! The 2nd generation Echo launched in 2017 re-imagined the first version to accommodate market influences with a new exterior finish to suit most people's home décor; a clever move by Amazon to make sure that everyone can enjoy the Echo's fantastic personal assistant capabilities.

The latest 2019 3rd generation Echo builds on this success with improved sound quality, a rounded design and a fabric covered finish available in 4 colors, including the rather fetching blue one (Amazon named it Twilight Blue, but personally I'd say it was more of an Ice Blue – it's pretty nice to look at either way!)

For me the choice of finish is an added bonus, what I like most about the Echo is how easy I have found it to adapt to my lifestyle. I use Echo all the time to:

- Set alarms and daily reminders
- Hear the weather forecast
- Listen to music

BEFORE WE GET STARTED: ECHO BASICS

- Get news flash briefings and traffic reports
- Speak to my other Echo-using friends and family hands-free
- Make use of all the other benefits this personal assistant offers

One of the things I like best is that most of the information requested from Echo also shows up on the Alexa App (more later), so I can view it there immediately or later when I want to review it.

I also like that if I unplug and move the Echo, the device takes just a minute or two to reboot and reconnect to Wi-Fi.

A quick note on the wake word: Echo requires a wake word to activate your device. For example, "Alexa, what's the weather like today" where "Alexa" is your wake word. I'm guessing you already know that, so in this guide I don't include a wake word when suggesting what to say to Alexa. For any voice command you read in this guide be sure to use your wake word first.

The Echo Preview

Getting started with Echo takes just a few minutes, and most of the setup happens automatically. Before you plug it in, though, look over your new Echo to familiarize yourself with the physical features of the hardware and the purpose of each.

9

There are four buttons on the top of the device:

Action Button: this is the button with the single dot; you can use it to activate Alexa without using the wake word, turn off timers and alarm. It is also required when you're setting up Wi-Fi.

Mic Off Button: this is the button with the image of microphone with the image of circle with a line drawn through it. Press and quickly release this button to turn off the microphones so that Alexa won't "hear" the wake word. However, it is worth noting that even when this function is disabled, you can still communicate with the device using the Alexa Voice Remote, a handy tool that is sold separately and discussed in the next chapter.

Volume Buttons: The left button (-) turns the volume down; the right button (+) turns it up. You can also use your voice to control the volume by saying, "Alexa, louder" for example.

Light ring: The LED light ring circles the top of the Echo and gives a variety of indications:

- **Blue with spinning cyan:** Echo is starting up after being plugged in.
- Blue (flashing): Alexa is processing your request/ giving you an answer.
- **Red:** The microphones have been turned off.
- **Orange:** Echo is having Wi-Fi connectivity issues (See Troubleshooting for details on trouble connecting to Wi-Fi).
- **Pulsing yellow light:** A message or notification is waiting for you.
- **Pulsing green light:** Someone is calling you or "dropping in" on your device.
- **Purple:** "Do Not Disturb" is on.
- **No lights:** If Echo is on and the light ring is off, the device is ready and waiting.

Power supply and port: The power supply plugs into the back of the device near the bottom.

Audio port: There is a 3.5mm audio port next to the power port where you can connect headphones or speakers. They are not provided with the Echo.

Specifications of the Amazon Echo

In case you're wondering, here are the specs of the device in front of you.

Dimensions and Weight:

- Height: 5.8 inches
- Width/ Diameter: 3.9 inches
- Weight: 27.5 ounces

Available finishes: Charcoal Fabric, Heather Gray Fabric, Sandstone Fabric, Oak Finish, Walnut Finish

Microphones: 7-mic array

Audio: Built-in 2.5" (63.5 mm) woofer and 0.6" (16 mm) tweeter

Networking:

- Wireless
- 802.11a/b/g/n, Bluetooth wireless protocol
- Dual-band (2.4 GHz and 5 GHz)
- For security reasons, Echo does not connect to ad-hoc or peer-to-peer networks

Warranty: 1 year

Echo Hardware Accessories: at the time of writing there were no specific accessories for the Echo, but it will likely only be a matter of time before other manufacturers opportunities for cases and stands.

Note on the Echo sound: one of the much-touted features of the 3rd Generation Echo is its improved speaker which features Dolby processing for better sound quality (Amazon have integrated the neodymium drivers and 3" woofer from the Echo Plus to do this). I'm not a real music buff, so personally I felt that the sound quality on the 2nd Generation Echo was already pretty good, but I am — hopefully reliably — informed by my audiophile colleagues that the bass is stronger and the sound quality in clearer.

1: Let's Begin! — Echo Setup

For any voice command you read in this chapter, be sure to use your wake word ("Alexa", "Amazon", etc) first.

Setting up the Echo is largely an automated process once you plug in the device. Here's what should have come in the box:

- Echo
- Power adapter
- Echo Quick Start guide

CJ's Tips: My Echo has a permanent spot on my office desk. Perhaps try several locations to determine where you get maximum usage from your Echo, and of course you may find that you move it occasionally. Wherever you place the device, Amazon recommends it is not placed too close to walls and windows, so that indoor echoes and outdoor noises don't confuse poor Alexa!

OK, enough with the preliminaries. Let's get the Echo started. Have your Wi-Fi password and your Amazon login information available for connection and setup.

Download the Alexa App to your phone and/or tablet (or see 5: below if you don't have a mobile device)

Mobile device requirements are:

- FireOS 3.0 or higher
- Android 5.0 or higher
- iOS 9.0 or higher

The quickest method for downloading the app is to go to the app store on your mobile device, look up the Alexa App and install it.

Once the Alexa App is installed, you will be able to access it on your mobile device and at alexa.amazon.com on your PC or Mac. It's a good idea to locate **Settings** on the App, because they are referred to often.

Plug in/Turn on Echo: Plug the adapter into Echo and an outlet, and the device will turn on.

Connect your Echo to a Wi-Fi Network

Access the Alexa App on your mobile device, PC or Mac. Follow the prompts on the App page to choose the Wi-Fi network you select from a list of those available. You may be required to enter your Wi-Fi password at this point.

- If the App page does not automatically go to the Wi-Fi connection page, then scroll down the App menu and find Settings. Then *choose Echo > Set up a new Echo*. You will then be given the option of connecting to a Wi-Fi network.
- If your network isn't on the list, you can add it by selecting ***Add a Network*** or ***Rescan***.
- If you have been asked to enter a password, you will also be offered the option to save it to Amazon. This will help you if you go on to add more Echo devices to the same Wi-Fi network as the Wi-Fi password will automatically appear during Wi-Fi setup.
- Select ***Connect*** and your Echo should confirm its connection to your Wi-Fi network.

Connection trouble: If in the unlikely event that your Echo does not connect to Wi-Fi, try these troubleshooting and solution tips in this order:

- Make sure your Wi-Fi is on
- Use another Wi-Fi device such as a tablet or computer to ensure Wi-Fi is working
- Tap the network you want to connect to and select the ***Forget*** option to reset the connection

- Attempt to connect to the network
- Make sure your network password (different than your Amazon password) is correct, and retry it
- If you saved your Wi-Fi password to Amazon but later changed it, you'll need to update the information on Amazon
- Turn off other devices connected to Wi-Fi; they might be causing too much congestion on the network for the Echo to connect
- Move your Echo away from a baby monitor or microwave oven that might be causing interference
- Try to connect to your router's 5GHz Wi-Fi frequency, if one is available, since that band might be less congested

Intermittent Wi-Fi: Sometimes an interruption in Wi-Fi prevents a connection. If your Wi-Fi goes out for brief periods, try this:

- Turn off the network hardware (router and separate modem if there is one)
- Wait 20-30 seconds
- Turn the modem on
- Turn the router on, if separate
- Unplug the Echo, and plug it in again after three seconds
- Attempt the connection to Wi-Fi again

If nothing works, contact your Internet provider or router manufacturer. If your network has an administrator, they might be able to help too.

Pair mobile devices with Echo

I recommend taking this step now since you're in a Startup frame of mind. This is a separate function from downloading the App to your mobile device, PC or Mac. Since Echo is Bluetooth-enabled, you can use it as a standalone Bluetooth speaker to stream your favorite audio services from your tablet or phone. This is particularly important if you predominately use Google Play Music and iTunes on your mobiles.

- Go to your phone's or other device's Settings and select Bluetooth to ensure that Bluetooth is On.

- Say, "Pair my phone" or "pair my tablet," or go to **Settings** on the Alexa App, select Echo and then follow the prompts for **Bluetooth > Pair a New Device**, and pairing will occur.
- Alexa will say, "Connected to _____ phone".
- Your device's Bluetooth Settings should list your Echo when the two are paired.
- Say, "Cancel," to stop Bluetooth pairing if you want to exit the mode before connection, and say "Disconnect" when you want to disconnect your mobile device from the Echo.

CJ's Tips: You can only connect the Echo with one mobile device at a time. It will automatically connect to the most recently paired device. So to change to a different device by Bluetooth, you will need to first disconnect the current device that is paired, by saying, "Disconnect" and then connect the new device following the above instructions.

Make sure your mobile device isn't asleep when Alexa tries to search for it. Also, depending on your mobile device, your device might prompt you for permission to connect to Alexa…if it does, go ahead and confirm.

Access the Alexa App on your computer, and sign in (optional)

This used to be my preferred way to use the Alexa App, as I like the larger view. The App address for your computer is: alexa.amazon.com. However, in recent months, I have realized that the Alexa App is not identical on different platforms. That is to say the layout, interface and some of the features are different depending on whether you're looking at the App on a PC or on a mobile phone. Throughout this book, if necessary, I will point out differences in the interface, but I strongly recommend using your mobile phone or tablet as that is the most popular way of accessing the Alexa App, and currently the best way to see most of the App's features.

Create your Voice Profile (optional)

You will of course know that Alexa can recognize any voice that speaks to her, but did you know that you can teach Alexa to recognize your particular voice and use that feature to link to these services for a personalized Alexa experience? Voice profiles can be set for up to 10 people on most Alexa-enabled devices. Once you have a Voice Profile you can enjoy these enhanced features:

- **Flash Briefing**, Alexa will give you news updates that are linked to your personal flash briefing choices.
- **Prime Music Unlimited** family plan users, say "Play music" and Alexa will play music tailored to your specific Music Unlimited profile.
- **Voice Purchasing**, when you have a Voice Profile saved then you don't need to give Alexa your voice purchasing code to proceed with a purchase.
- **Calling & Messaging**, when using this feature Alexa will automatically call or message people listed only on your personal contacts.

To create your Voice Profile, simply ask Alexa to "Learn my voice", and then follow the vocal prompts from Alexa. You will be asked to repeat 10 phrases like "Alexa, order pink pajamas" and "Amazon, feed the cat". And then Alexa says, "Nice to meet you" when she's done with setting up your voice profile.

To create a Voice Profile for a different person, not listed on your account at Set Up, then tap on Settings and then scroll down to log out; then sign back in to your Alexa App account, where under your name you should see the option, *I'm someone else.* Click or tap on this and follow the instructions again to create a different personal profile. Then you can ask Alexa to "learn my voice" again, using the new profile.

Note that if you've already created an Amazon Household (see below) then members of the Household can say "Switch accounts" and then ask Alexa to "Learn my voice".

To delete a Voice Profile, select **Settings > Accounts > Your Voice > Forget my voice** and confirm.

Customize the Echo device sounds

I suggest you wait a few days or a week before customizing the Echo to see how you like default mode, but the information is included here as part of the general setup process.

- Go to **Settings** on the Alexa App
- Tap **Sounds** and choose levels for Media volume and Alarm, Timer and Notification Volume before browsing Custom Sounds.

Create an Amazon Household

This is another setup feature you might want to wait till later to do, if at all.

If there are other Amazonians in your household — your literal or figurative "household" — you might want to share capabilities such as being able to listen to each other's music. You can explore the benefits and how-to's of Amazon Households in more detail by following the instructions below or going to www.amazon.com/myh/manage.

- On Amazon.com hover over **Accounts & Lists** near the top of the page
- Select "Your Account"
- In the "Shopping programs and rentals" box (bottom right), select **Amazon Households** to go to the Manage Your Household / Your Amazon Household Benefits page
- Select "Add an Adult," and follow the instructions including providing their login information, so they will have to give it to you or be there with you to type it in
- And/or select "Add a Child," Create and Save their Profile, and follow the link to **Manage Your Content and Devices** to determine the content they can access

Set up the Alexa voice remote (optional)

The Alexa Voice Remote is available on Amazon for $29.99. Its availability has been intermittent, though at this writing, it is in stock. Be aware that there is also an Alexa remote controller for Amazon Fire TV and Fire TV Stick. That's not what you want here.

The Voice Remote is relatively useful since, while Echo is equipped with Far-field voice recognition to pick up the wake word and queries, there are times the remote is a handy alternative to talking across the room, especially when there is other noise in the space that will cause you to shout or that will confuse Alexa.

Once the remote's batteries are installed, it might pair automatically. This has been hit and miss with users. If it doesn't, you can pair it via the Alexa App.

Pair the remote using the app by going to **Settings**; select your device (your Echo); select **Pair Remote**.

CJ's Tips: I'm not a fan of the remote because I often pair my phone with my Echo. You can only pair one Bluetooth device at a time so I choose my phone over a voice remote!

2: Meet the Alexa App

*For any voice command you read in this chapter,
be sure to use your wake word ("Alexa", "Amazon", etc) first.*

Now that Setup is complete, you'll get the most utility from the Echo and complete tasks more quickly if you know your way around the Alexa App. We will refer to **Settings** several times here. They are a part of the App, but there's so much information to share about **Settings** that it has its own chapter (page 97). It's inevitable that some information will be covered in both places.

Amazon calls Alexa the "brain behind" its Echo devices. Alexa is cloud-based, making it possible for Alexa and Echo to continually be improved and updated. If you're worried about security, consider that it's likely that many of your accounts are already somewhere in the cloud — banking, credit cards, medical in addition to other Amazon accounts such as Prime or Drive. The same level of security used with that information is used with Alexa.

The App and your Voice

Most of what you can do via the Alexa App can also be done with voice. If you're listening to an Amazon music station, for example, you can select the Pause button on the App or simply say, "pause." In the remainder of this guide, we will note things you cannot do by voice, such as Change the Wake Word. In all other cases, assume you can accomplish the task with your voice, and give it a try.

There is also interplay between the App and the Echo. For example, when you're listening to media, information about what your listening to will display on the **Now Playing** page of the App, or when you add an item to your **Shopping List**, you can consult the the list on the relevant App page too.

A Word about Help & Feedback

The *Help & Feedback* page, accessible within the Alexa App, is discussed in detail later, but it's worth knowing before we get started that this section contains a wealth of information about all the topics discussed throughout the rest of this book.

So let's start at the beginning and select the Homepage of the App.

A Word About the Alexa App Layout

s you view the App layout on a PC or Mac, you'll see a Menu of App pages on the left and the current page on the right. That's what you might see on large mobile devices too, but on most phones and tablets, you will see just the current page; the Menu will be depicted with three horizontal lines at the top left. You will need to tap on this Menu icon to navigate to different pages of the App.

Note that on the tablet and mobile phone version of the App; no matter what page you are on, there are four icons at the bottom of the page – a *Home* icon to take you back to the *Homepage* which is a trapezoid shape with two lines under it, I think it's supposed to represent the Cards (more about those later); a Communications icon (a little speech bubble) to take you to the relevant pages of the App for making calls and sending messages; a Play icon – a circle with a triangle inside - which takes you to the *Music & Books* pages; and a Devices icon (a little house with two switches inside) which takes you to the Devices page where you can access and manage the Amazon Echo devices and any other smart home devices that are integrated into your Alexa system (discussed further later on).

So, let's start at the beginning and select the *Homepage* of the App.

Alexa App Homepage on your PC or Mac

When you're on the *Homepage*, starting at the top, what you'll see are different sections which are Things to Try, your Cards and, when you're playing most media, a media Player at the bottom of the page will appear.

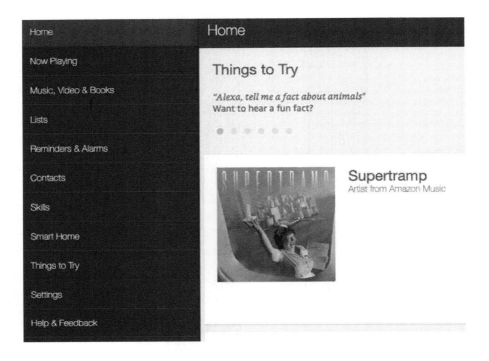

Things to Try: When on the **Homepage**, a **Things to Try** feature appears at the very top. It rotates through some suggestions of things you can say to Alexa related to current or trending issues.

There are shaded buttons below the suggestions. I click manually through them since I don't have the patience to watch them rotate at the rate of 7–8 seconds per suggestion! Occasionally something of interest appears, like *Brain Facts — Learn something new about the human brain* — but I've found they can be a time waster in the same way social media can be when there are other things I should be doing.

Cards: The next thing you'll see after the Things to Try suggestions are your Cards. A Card is created with most verbal interactions with Alexa. The Card preserves the request and response from Alexa such as a music station or a forecast you requested, a Skill used or items you added to your shopping list. Often these cards aren't of much interest, but as you'll discover later there are times when they are, like after you've asked for weather, movie or local business information.

Cards form your Alexa Dialog History. You can scroll down the Cards to review them. A list of them is also available in *Settings*. Select *Settings*, and scroll down to the *General* section, and select *History*.

Card options: On the right of each Card you'll see the word More with a V-shaped icon, known as a caret. Click on it to see options for further action:

- **Remove card:** Select this to delete the card
- **Give Voice feedback:** The Card will show what "Alexa heard" with the question, "Did Alexa do what you wanted?" Answering the question helps Alexa improve its understanding of how you speak. You don't have to answer the question, of course. If you answer "no," you'll be given the option to "Send more detailed feedback". Select that for unlimited space to provide feedback and the opportunity to request that Amazon get back to you on your input.
- **Learn More:** This takes you to Alexa and Alexa Device FAQs, which is a useful part of the *Help & Feedback* section of the menu I recommend you browse when you have a few minutes.

To use these options on older Cards, select *More*, and the box will expand to show them.

Deleting your entire history at once is only possible by deregistering your device, something you'd likely do only if giving it away or selling it. Deregistering the Echo and reregistering it is explained in the *Settings* section.

Bing Search and Wikipedia: One or both these options might appear on a Card:

- **Search Bing:** If Alexa is unclear about what you say or if you ask about a person, place or thing, Alexa will provide the option to "Search Bing for…" whatever your query was about. Bing is your only search engine option on the App.
- **Learn more on Wikipedia:** Depending on the subject you ask Alexa about, the Card might give you a short answer and a link to the topic on Wikipedia.

Player: When media (songs, radio, podcasts, etc.) is loaded, whether playing or paused, the Player appears at the bottom of the Alexa App page and allows you to control it. The bar includes an icon of the current song in the queue. The Player appears at the bottom of the App regardless of what page you're on. It is discussed fully in the ***Now Playing*** section next.

The Alexa App Homepage on your mobile device

As mentioned earlier the Alexa App interface is quite different on a mobile compared to your PC or Mac; particularly the Homepage of the App on your mobile device looks very different.

What you'll see on your mobile device at the top are three horizontal bars on the left hand corner (when you tap on this you'll get a ***Menu*** of other options); a question mark icon that leads to the **Help & Feedback** section on the right hand corner; and a greeting (Good Morning, Good Afternoon etc...) in the middle.

Below the greeting you'll be able to scroll upwards through a series of floating tabs that either take you to different pages of the App, offer information on new Skills or Things to Try, or – once you've started using Alexa – will show links to where you left off on you Audible book or the Music currently playing/ or last played. These tabs act a bit like the Cards mentioned above.

However, to get the full options for your Cards via the App on your mobile device, then go to **Menu > Activity**.

As with the App on your PC or Mac, when you play any media via your Echo, the **Player** will appear at the bottom of your current page.

3: The 'Now Playing' Page and Player

For any voice command you read in this chapter, be sure to use your wake word ("Alexa", "Amazon", etc) first.

One of Alexa's most attractive features is its ability to play media from a wide range of sources.

I appreciate that you might not yet have anything to play yet and therefore much of this chapter might not make sense to you. If this is the case feel free to skip ahead and return to this chapter when you are ready.

Now, as we mentioned earlier, on some pages of the Alexa App there can be significant differences in the layout and interface when you view it on a tablet or mobile phone compared to a PC or Mac. The **Now Playing** page is one of these pages where it is quite different.

Basically it doesn't exist as an option in the **Menu** on the tablet or mobile phone version of the Alexa App!

To access the **Now Playing** page on your tablet or mobile phone, when you have media playing you need to then tap on the **Player**, which loads on the bottom of whatever page you're viewing (the **Player** is always visible on every page of the mobile App when media plays).

To access the **Now Playing** page on your PC or Mac, simply click on the option from the Menu.

Whichever way you get to the **Now Playing** page, the same options will be offered, but in slightly different, not particularly significant, layouts.

For much of your audio content, the **Now Playing** page displays an image of the media playing – for example an album cover image –

plus an icon for the source such as iHeartRadio, Kindle audiobooks or a podcast on Tunein to remind you where the media is coming from. Under those images, is the media **Player**.

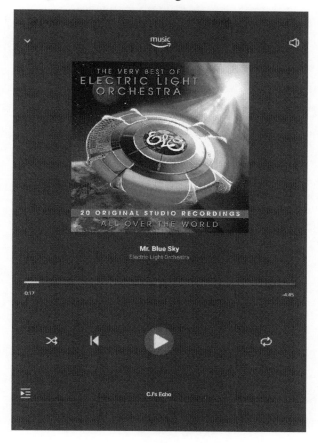

You can manually tap or click any option on the **Now Playing** page or you can use voice. The **Player** always offers these control symbols:

- **Play/Pause:** Manually select or say, "Pause," "Resume," "Play" or "Stop."
- **Go back:** Manually select or say, "Go back."
- **Go forward:** Manually select or say, "Go forward."
- **Sound level vertical slide:** Slide the volume bar manually by "grabbing" it or use a verbal command such as "volume seven" or "mute." Note that on your tablet or mobile phone, the Loudspeaker

icon may be at the top right of the Now Playing page.

- **Song progress bar:** When using the App on a computer, selecting a place earlier or later in the song isn't possible. Ability to change the place in the song varies on mobile devices.

Depending on what type of audio you are playing, additional options are shown on the player:

- **Thumb Up & Thumb Down:** Rate the song, and Alexa will use your opinion to tailor your music stations to your preferences
- **Shuffle:** Selecting the crossed arrows will cause the Queue to be played in random order
- **Repeat:** Select the looping arrows to repeat the song or station when it is completed

Also visible on the *Now Playing* page on a PC or Mac, are two lists you can view, Queue and *History* by selecting your choice.

Queue: This is the list of selections in the media you've chosen such as the Classical Focus Prime Station I'm currently enjoying. Expand your options for each piece by selecting the V-shaped caret. For music, expanding the box allows you to choose further actions like:

- Add the song to your Library (we will go into much more detail about your music library in Chapter 5, page 34)
- Rate the song by choosing Thumb Up or Thumb Down.
- Shop the Digital Music Store, which will open in a new page with the standard features seen on all Amazon product pages.
- View album in Prime Music, if applicable, which will open in a page that looks very much like the Digital Store page but with the opportunity to buy any/all the songs on the recording.
- When listening to podcasts or radio stations, clicking the caret and expanding the box allows you to mark what you're listening to as a Favorite program.

History: This is the list of media you've played from all sources.

Play anything from Queue or History by tapping/clicking it. Your choice will then become the audio content that is **Now Playing**.

When viewing the **Now Playing** page on a tablet or mobile phone, you will see an icon with 4 lines and a triangle; tap on this to access the **Queue** (if there is one), where you can then change what you're listening to if you wish. Other options as discussed above are more limited, or in fact not available, on the mobile version of the App.

Note that if you want to see your media **History** on a tablet or mobile, then select the Play icon at the bottom of the page (the circle with a triangle inside) and the top of the page will show your most recently played media..

CJ's Tips: Note that if you are playing a station and then select another media, the station might be canceled. I learned this by pausing a station and asking for my news **Flash Briefing** (explained later but found in **Settings** if you want to explore it before we get to it). The station did not resume, even at my request. Stations are not canceled if you ask for your weather forecast, something from Wikipedia or a simple question. The station volume is reduced while Alexa fulfills the request. I'm still learning which requests cancel stations and which don't. If a station gets canceled, and I want to continue listening, it's right there in History where I can pick up where I left off.

As mentioned above, if any of this is confusing right now then read on and return to this chapter once you have started playing some audio.

4: Watch This — Video

*For any voice command you read in this chapter,
be sure to use your wake word ("Alexa", "Amazon", etc) first.*

As the standard Echo does not have a screen, you might feel this section is irrelevant to you, but it's worth reviewing what video services are available via the Alexa App and how to access them as some of them can be used on your home TV set and controlled via Echo with your voice.

Before we discuss all the Video options on your Echo, please note that during the writing of this book, there was a major update to the Alexa App the result of which is that the **Video** section is no longer visible on the App when viewed via a tablet or mobile phone. This situation may well change with the next update.

At the time of writing I could still access the Video section via the App on my PC and Mac, go to the **Menu** and select **Music, Video & Books** and Video services are the first section on the page.

Also, it's important to understand that there are two types of video services that you can use with Echo devices.

The first is what I call direct access, meaning that if you have an account with these services you can watch their content directly on Echo devices that have a screen such as the Echo Show. This is the case for Amazon Video and Prime Video, IMDB movie trailers, Hulu and NBC.

The second is best termed as indirect access; this is where you use your Alexa voice commands via any connected Alexa-enabled device, such as the Echo, to control the content streamed to your home television set from streaming or cable services such as Fire TV or Dish TV.

So, for your Echo you will have the possibility of setting up the following indirect access TV services:

Fire TV

The original Fire TV dongle from Amazon is a comprehensive media-streaming device. It can be installed and used entirely independently, with or without its own Alexa-enabled remote control. Or, it can also be paired with and controlled by other Alexa-enabled devices, such as the Amazon Echo.

The same is true for the Toshiba Fire TV edition television set and the – a smart TV that has Fire TV streaming already integrated and which can be voice-controlled when you pair it to the Alexa-enable remote or to one of your other Alexa-enable devices.

Furthermore, there is now also the Fire TV Cube, which already has Alexa preinstalled, ready to operate with voice control alone.

What you can watch with Fire TV: Most streaming and live TV services can be viewed with Fire TV including: Netflix, Hulu, HBO Now, Crackle, Amazon, ESPN, Showtime, Sling, DirectTV Now, NBA, MLB. TV, CNN, Comedy Central, HGTV and AMC.

What else you can do with Fire TV and Alexa: Enjoy music services played through your equipment such as Amazon Music, and iHeartRadio, order pizza or an Uber, browse Yelp and play games from developers like EA and Disney. Search "Fire TV" on Amazon to see everything the service offers.

Connecting Alexa and Fire TV: Amazon provides complete instructions for setting up Fire TV equipment and a PDF Users Guide to make the most of Fire TV. To control Fire TV with Alexa:

- Go to ***Music, Video & Books*** in the Alexa App, and select Fire TV
- Select ***Link Your Alexa Device***
- Select the device you want to link
- Follow the onscreen prompts to complete linking of Alexa to the Fire TV Player

Once setup is complete, put Alexa to work with requests like:

- "Watch House of Cards"

- "Next episode"
- "Rewind 10 minutes"
- "Jump to 30 minutes"
- "Show me Emma Watson movies"
- "Play Sia music"
- "Order Domino's pizza"

And literally thousands of other requests. To be clear, linking your Echo will allow you to use Alexa to control your Fire TV options on your TV, and will only work to watch Fire TV content on your television screen if you have one of the Fire TV devices.

Dish TV

This service appeals mostly to those that already have a Dish TV package, which currently starts at $59.99/month and includes the Hopper DVR. Dish TV pushes its Alexa relationship by periodically offering new customers an Echo device upon signup. You can check the availability of this offer at www.dish.com.

To be clear, to control Dish TV via Alexa you will need a Dish TV Package and the DISH Hopper Smart DVR (the Hopper 3 and newer, Hopper with Sling, Hopper and Wally are all currently supported.)

Connecting Dish TV to Alexa:

- Go to *Music, Videos & Books*
- Select *Dish TV*
- Enable the Skill
- Have your Dish TV and Amazon login information handy, since you might need to enter it during setup
- Follow the on-screen instructions to connect Alexa to Dish TV which include 1.) turning on your Hopper setup box and TV 2.) Enter the code given on the TV screen into the Alexa App

- You might need to update your Hopper's software to the latest version before you can use Dish TV with Alexa. On your Dish set-top box, navigate to channel 9607 and locate/select "software update" to update your receiver to the latest software. Allow the software update to complete and the Hopper box to restart before returning to the Alexa app to attempt to link again
- Select **Finish Setup** in the Alexa app
- Follow any on-screen prompts to link Echo to the TV or other equipment you use to watch TV and other video services

If you have issues with the process, try these fixes:

Update the Alexa App, and make sure you signed into the App with the same account information you used to register your Alexa-enabled device - that is, make sure your app and device are on the same account. Double-check to make sure you've enabled the Alexa Video Skill for Dish TV (or another video provider you're setting up). And failing all of that, contact Dish TV directly.

Once setup is complete you can ask Alexa to control your Dish TV viewing by saying things like:

"Go to channel 75"

"Find Modern Family on Dish" — Alexa will look for options on the services you have such as Netflix, Hulu and Amazon Prime)

Play your recorded content ("Play Dallas Cowboys football game")

Pause, fast-forward and rewind what's being shown ("Rewind 2 minutes")

5: Songs by the Millions — Music

*For any voice command you read in this chapter,
be sure to use your wake word ("Alexa", "Amazon", etc) first.*

It's difficult to beat Alexa for ease and convenience for playing music from a diverse range of sources. But working out how to find and set up these different sources is not always that obvious! For a start, from the Alexa App on a tablet or mobile phone, accessing the different Music services is quite different compared to what you see on a PC or Mac.

On a tablet or a mobile phone there is a round icon with a triangle inside, the Play icon, which takes you to the **Music & Books** page.

On this page you will see any Amazon Music, Audible and Kindle ebooks that you already own and have recently played/ viewed. This area will be blank if you haven't used any of those media yet. For example, when I first started my Alexa App on my tablet, I'd already downloaded some Kindle books to that tablet, so that's what I saw on the landing page; when I then set up an Audible and Amazon Prime Music account, I subsequently found my music and audio books listed here too.

Next, near the top of the **Music & Books** landing page you will see the word **Browse** – tap on this to see the different sources of Music that can be connected to your Alexa, such as Amazon My Music, Amazon Prime Music, Spotify, TuneIn and so on.

Finally, on a tablet, as you scroll down the page you will see a number of suggestions of things you might like to do, for example, "Link to your favorite streaming services", click on those suggestions to explore the option further.

From the Alexa App on a PC or Mac, when you select **Menu > Music, Video & Books**, you will simply see the list of streaming options available in each category: My Music Library, Amazon Prime Music, Spotify, TuneIn etc.

So now let's look at the Music services you can connect to via Alexa in more detail:

My Music Library/ My Amazon Music

On a PC or Mac, you will clearly see **My Music Library** labeled on the first page of the Music, Books & Videos.

On a tablet or mobile phone, after selecting **Music & Books > Browse Music** you will see the label **My Amazon Music**. Tap on this to see **My Music Library**.

To be clear, this section of the App is where you access YOUR personal music collection; the next section, Amazon Prime Music, is where you can access ALL music available from Amazon.

The order seemed logically backwards to me when **My Music Library** was sparse, but now that I have a large Library of music, it makes sense. I access what I already have more often than I want to browse what else is available.

Your account's My Music Library will be populated with music that has been:

- Purchased from Amazon.
- Selected in the form of a Playlist from Amazon Prime and Amazon Music Unlimited subscriptions.

CJ's Tips: The Amazon Music app (https://amzn.to/2ElWBci) for your desktop computer is Amazon's equivalent of the iTunes app. Once you have it installed you can play all the music in your Amazon Music account on your PC or Mac just like you probably do on iTunes already.

So, now let's explore the **My Music Library** further.

Firstly, right under the words My Music Library, you may be asked to select the device you want to play music on (only relevant if you have more than one Alexa-enabled device.)

You may also have the option to select the music Library you want to access. If you have set up an Amazon Household, you can access Libraries of household members too. An Amazon Household can include up to two adults, though each must have an Amazon account, and up to four children. Creating and managing an Amazon Household is done through your Amazon account. Here again are brief instructions and options for Amazon Households:

1. Hover over **Accounts & Lists** near the top of the page on Amazon.com
2. Select **Your Account**.
3. In the **Shopping programs and rentals** box, select **Amazon Households** to go to the **Manage Your Household / Your Amazon Household Benefits** page.
4. Select **Add an Adult**, and follow the instructions including providing their login information, so they will have to give it to you or be there with you to type it in.
5. And/or select **Add a Child**, Create and Save their Profile, and follow the link to Manage Your Content and Devices to determine the

content they can access.

Here is a direct link: www.amazon.com/myh/manage

Search Your Library

Use the search box to find songs or albums more quickly in your Library. As you type the search word, a list of results will begin to populate and gradually narrow the more of the term you complete.

The next row is a menu of tabs: **Playlists/ Artists/ Albums/ Songs/ Genres**. Using them is a very straightforward process, so just a few comments will suffice for each.

Playlists: There are three sections here:

Auto Playlists: Purchased — a list of all the songs you've purchased from Amazon, which in time might become a very long and eclectic list; and Recently Added — which includes recently imported playlists and a list of songs purchased in the last month or so.

Your Playlists: These are playlists that you can create within the Amazon Prime Music service.

Prime Playlists: These are preselected playlists that are available on Amazon Prime Music. If you don't have an Amazon Prime membership, then you won't see anything here. If you do, then it's time to add some of these Prime Playlists to your Music Library (see instructions below)!

Artists/ Albums/ Songs/ Genres: These tabs all function the same, and you can find the music you want in two ways: scroll down through the alphabetical list at the left or jump down to its location using the #ABCDEFG, etc., list. Of course, you can also just ask Alexa to play the music you want.

CJ Tips: In my experience, it helps to be as specific as possible when asking for music, such as, "Play Bruce Springsteen Born to Run from My Music Library," or "Play Bruce Springsteen Born to Run from Amazon Prime Music." Once you have a few different music services set up it helps Alexa if you let her know exactly which service you want the music to come from. I will talk about setting Default Music Services later to make this process even easier.

Amazon Prime Music

On a PC or Mac, you will clearly see **Prime Music** labeled on the first page of the Music, Books & Videos, under *My Music Library*.

On a tablet or mobile phone, after selecting *Music & Books > Browse Music* you will see the icon for Prime Music, which is the word music in lower case with the Amazon swoosh arrow underneath.

For this section to be relevant, you'll need a subscription to Amazon Prime and/or Amazon Music Unlimited.

Prime Music: This option is available as part of Amazon's standard monthly Prime membership deal, to be clear, you need to have the Prime membership deal to benefit.

- **Benefits:** stream 2 million songs in Prime Music and a long list of shipping, shopping and media benefits that can be viewed on Amazon.
- **Standard annual:** $119/year, after 30-day free trial.
- *Standard monthly:* $12.99/month, after a 30-day free trial.

There is also a student plan for $6.49/month or $59/year; and a plan for those who qualify for an EBT or Medicaid card, $5.99/ month.

Amazon Music Unlimited: If you do not wish to have a full Prime membership, then you can still access Amazon Music with this offer.

- "Unlimited access to tens of millions of songs" according to Amazon.
- New members get the first month free and then switch to a paid-for plan; Individual plans are currently $7.99/ month, whilst a family plan is $14.99/month which can be used by up to 6 members. Annual memberships both for individuals and families are also available.
- There is also a specific single device plan for use with an Amazon Echo or a Fire TV device for $3.99/ month. And a student plan for

just $4.99/ month.

CJ's Tips: For what it's worth, my recommendation is getting a Prime membership since it offers so many other benefits, and if you then don't find enough of the kind of music you enjoy, consider adding a Music Unlimited subscription. Again, you can manage all music subscriptions from this here: www.amazon.com/music/settings

Amazon Music Unlimited HD: Launched in September 2019 to coincide with the launch of the new high-fidelity smart speaker, Echo Studio. This is a further tier of music provision from Amazon with over 50 million songs available in High Definition (HD) or Ultra High Definition (ultra HD). If you are a serious music buff, this is an option with considering. New users get 90 days free and then switch to a paid-for plan. At the time of writing the payment options available were:

- Amazon Music Unlimited subscribers: an additional $5 per month (Individual or Family plans only)
- Prime members: an additional $12.99 per month
- General Amazon customer (neither Prime members nor Amazon Music Unlimited subscriber): $14.99 per month

Search your Amazon Prime Music

Once you have your Prime music subscription you can return to the Alexa App and explore. The options are labeled in two categories, Stations and Playlists.

Stations: Amazon's Stations are music streams based on either a particular artist, era of music, or genre of music. Theoretically, if you choose a Station – just like the radio in the olden days – you will get that type of music on an endless stream. This is a good way to search for music in line with your preferences.

- **All Stations:** This section has two categories to browse, and when you make a selection, the Station begins to play. Popular Genres & Artists stations include Classical Focus, Top Pop, Smooth Jazz,

Lullabies and other broad categories. Each of the All Artists (A-Z) stations bring together the very best music from across the band's or artist's career. Try this feature for yourself, of course, but I rarely use it. There is no search function, and scrolling down thousands of artists is impractical.

- **Genres:** This is a more reasonable way to locate music you like, but you must go two levels deep to choose stations. Search categories including Featured, Pop, Country, Classic Rock, Christian, Alternative, Reggae, Clean, World, R&B, Classical and many more – even Christmas! Choose a genre and, you are given the list of subgenre "stations". For example, choosing the Classical genre gives you the choice of dozens of stations like Baroque, Classical Piano, Classical Piano, etc., and popular artists like Yo-Yo Ma, the Emerson String Quartet and Hilary Hahn.

Playlists: Amazon's Playlists are curated "albums" of music that have been put together by Amazon's team. Like the stations they are categorized according to genre or music era or artist, but in this case there will be a finite selection of tracks – for example 50 Great Feel-Good Classics. More than 1,700 Amazon Music Playlists have been curated, and the number is growing, so give yourself a day off to have a good look around!

You can simply click on a Station or a Playlist shown on the Alexa App pages to have Alexa play it; or when you know the title of the Playlist, you can just ask Alexa, for example: "Play 90s Dance Anthems".

However, to avoid too much searching around, you might prefer to add the Prime Playlists that you like to your Music Library.

Add Prime Playlists to My Music Library

There are two ways to add Prime Playlists:

Manually via Amazon.com:

- Go to your Amazon Prime Music online (music.amazon.com)
- Browse Playlists by clicking on the link at the top of the left hand side menu and continue finessing your options until you find the

Playlist that interests you.

- Hover or tap a Playlist to see your options:
- Select the **+ *sign*** to add the entire list to your My Music Library "sight unseen," or later the X sign to remove it.
- Choose the Play icon to listen to the list, or later, the Pause icon to pause it; and then add it if you like it as above.
- Select the three dots to:
  - Share the list: Copy the link, copy the page code to embed elsewhere,
  - email the link or share it on Facebook or Twitter.
  - Open the Playlist to view how long it is, number of songs in the list, who curated the list and what its reviews are. There, you can also just choose individual songs form the list to add to your Music Library.

By Voice via the Alexa App:

- First go into the Alexa App to **Music & Books > Amazon Prime Music > Playlists** and find a Playlist you want to add.
- Click/tap the playlist to start it playing. Once it's playing you can then ask "Alexa, add this playlist to my library" and Alexa will do just that!

Once you've added Prime Playlists to your Music Library, when you next go into the Alexa App you'll see the Prime Playlists listed in the My Music Library pages.

CJ's Tips: Every track you hear on a Prime Playlist or Station comes from an album that is also available for you to listen to free as a Prime member. However, adding an album to your Music Library may require a little bit of detective work.

Here's an example... you hear the song "Don't" by Ed Sheeran on a Prime Playlist and decide you would like to listen to the whole album that the song comes from.

- First go online to Amazon Prime Music by going to music.amazon.com in the browser of your desktop computer.
- Second search "Don't Ed Sheeran" in the search box at the top of the page and hit enter.
- Third, your results will be shown so scroll down to Prime Songs and when you see the song "Don't, Ed Sheeran" click the three vertical dots to the right of the track. A pop up window will appear giving you further options and one of those options should be **View Album**.
- Fourth, go ahead and click **View Album** and you will be shown the entire album that the track is from **plus** the option to + **Add to My Music**, and clicking that will add the album (in this example the album is called "X (Wembley Edition)") to your music library.

Now you're all set and can just ask "Alexa, play the album X from my Music Library"

Spotify

With 232 million users and 108 million subscribers, www.spotify.com is one of the largest music streaming platforms. While Spotify has a free subscription, a premium account is required for use with Echo. That's Spotify's choice. The current cost of a premium account is $9.99/month for those who have not tried Premium before. Besides using their service with the Echo, benefits include the ability to download music to your devices to listen offline, skip songs you don't like as often as you want, play any song and listen without ads.

To create an account, choose an email address to link to, a password, user name and payment method, which can be a credit card or PayPal. If you choose PayPal, Spotify will connect to PayPal, and you might have to sign in. Once on PayPal, you will select which payment associated with your account you want to use.

Linking Alexa to Spotify: Once you have a premium account, select **Spotify** from the **Music, Video & Books** tab, and choose: **Link**

your account* > *Authorize the Account (If asked).

A new window will open where you will have to authorize the connection of Alexa to your Spotify account. Once you do that, return to the Alexa App where you will be asked whether you want to make Spotify your default music service. If you want to change that later, it can be done at ***Settings > Music & Media > Choose default music services***. There, you can select a Default music library and a Default station service.

If you want to unlink your Spotify account, for example if you no longer own the Echo, select ***Settings > Music & Media > Spotify*** and ***Unlink Account from Alexa***. On that page, you can also go to Spotify to Manage Spotify Settings such as upgrade to a Premium for Family account, view and edit your profile or see when the next Renew date is.

Using Spotify on Echo: If you've made Spotify your Default music service, you can use your voice or the Spotify app to play music from the service. For voice, simply request music from your favorite artist, and Alexa will play it. If it's not the Default service, you'll have to say, "Play Selena Gomez on Spotify," for example. As usual, the ***Player*** will appear on the ***Now Playing*** page and at the bottom of any other page you're viewing. Manually control Player functions using the Alexa App, with your voice – Play, Skip, Shuffle, etc.

Pandora

Pandora is often referred to as an online radio pioneer, having been launched at the dawn of Internet radio streaming services back in 2000. Currently Pandora is only available in the USA. But its popularity is as good as ever, with over 160 million users, and over 30 million tracks available, it's safe to say that Pandora will be here for a while longer.

One of the benefits of Pandora, for our purposes, is that you can use Alexa with a free Pandora account, which I do. If you already have an account, you can skip to Linking Alexa to Pandora below.

There are currently three membership levels, see www.pandora.com:

- **Pandora Free:** Free; Ad-supported radio; Personalized stations.
- **Pandora Plus:** $4.99/month; 30-day free trial; Ad-free Personalized stations; Better audio; Offline radio.
- **Pandora Premium:** $9.99/month; 60-day free trial; Ad-free; Create new stations and playlists rather than personalizing existing stations; Better audio; No limit on skips and replays; Download music to other devices.

Select the level you prefer, and you'll be taken to an account signup page. Even if you select a free trial for a paid subscription, providing an email address and password on the first page will immediately begin a free account.

Linking Alexa to Pandora: Back on the Alexa App, select *Pandora* from the *Music, Video & Books* page. Then, choose: *Link your account > Authorize the link*

Using Pandora on Echo: Once Alexa and Pandora are linked, you'll have two options on the Pandora page on the Alexa App.

- Choose + **Create Station**, and enter an artist, genre or track to search or scroll through the **Browse Genre** section to view more than 60 genres from standards like Rock, Alternative, Classical, Hip Hop/Rap, Instrumental, Comedy and Decades to niche categories like Game Day, Mexico, Rainy Day, Pandora Local and Driving.
- Browse and choose one of the stations listed below **My Stations**, a list that will populate once you start adding stations or will be displayed immediately if you link to an existing account. Toggle Sort by date (date added) and Sort by A-Z to view the My Stations list. The first entry will always be Shuffle to mix the list randomly.

As usual, the *Player* will appear and at the bottom of the page and on the *Now Playing* page and can be used to control the music. Voice controls work too.

CJ's Tips: I like Pandora, they have some great stations and I have no problem asking Alexa for a station based around an artist like, "Alexa, play Adele on Pandora," but things get more problematic when I ask for a themed playlist like "Alexa, play Laid Back Brunch on Pandora." There were many stations like this that got no response from Alexa even when I set Pandora as my default station service.

If you encounter this problem then the solution is simply to go to **Music, Video & Books > *Pandora* > *Browse Genre*** on the Alexa App and start a playlist playing. It will immediately be added as one of your stations and will appear under My Stations. Now try again to ask for it with your voice and you should have no problem.

iHeartRadio

This radio network and internet radio platform offers an immense amount of content including music and podcasts provided by 800 iHeartRadio partner stations in the US, more than 1,000 artist stations and a range of other media.

There are three membership levels:

- **iHeartRadio:** No cost, but you're limited to choosing a local radio station or curated stations built around well-known artists but including similar artists.
- **iHeartRadio Plus:** $4.99/month through Alexa; Play any song on demand; Unlimited skips; Save songs to create a single playlist; Replay songs from live radio and custom stations.
- **iHeartRadio All Access:** $9.99 through Alexa; Everything offered in Plus; Unlimited access to millions of songs; Listen offline on iOS and Android versions, but not yet on Alexa; Create as many playlists as you like.

I tried All Access but settled on the free iHeartRadio account simply because I've got so many other music choices that I don't need a paid service. If you start out with a free account, select the Upgrade Now tab on iHeartRadio at any time to try one of the paid plans.

Linking iHeartRadio to Alexa: Select **iHeartRadio** from the ***Music, Video & Books*** page and then ***Link Your Account.***

Authorize the link to an existing account or create a free or upgraded account with personal and payment information.

Using iHeartRadio on Echo: The homepage has a search box where you can type in keywords including an artist's name, genre or city, for example. Searches return results related to your keyword, and they're divided into four categories:

- **Stations** — Artist Stations are titled after well-known artists to give you an idea of the flavor of the music, but each station includes songs from a range of artists. I'm currently listening to the station called Adele. The song Queue includes Rihanna, Ed Sheeran, Lorde, Ellie Goulding, OneRepublic and others in addition to Adele.
- **Songs** — Singles by title or group name related to the keyword.
- **Artists** — Bands, solo acts, choirs, etc. with names related to the keyword.
- **Talk Shows** — Shows and podcasts with names related to the keyword.

Your second option to locate music is the Browse section with three categories to look through:

- **Live Radio:** 800+ radio stations from around the country.
- **Shows:** These are mostly podcasts and the 20+ categories include Business & Finance, Comedy, Crime, Entertainment, Politics, Spirituality and Sports. Categories contain a dozen to more than 100 show options.
- **Favorites:** You have the option to select any station or show as a Favorite, and it will be saved here for easy locating later, though it might not show up immediately.

Manually select any Station and it will begin to play. You can also try using voice, but be as specific as possible for the most accurate results.

Once playing, the content can be controlled on the ***Now Playing*** page or with your voice. Also on the ***Now Playing*** page, you have the option to expand any selection in the ***Queue*** to:

- Tune the Station by choosing what you want to play — Top Hits, Mix or Variety. I don't know what the difference is between Mix and Variety, though perhaps seasoned iHeart enthusiasts will.
- Select the Station as a Favorite (check out my Tip below).
- Rate the song with thumb up or down symbols.
- Create a new station around your favorite artists.
- Shop Digital Music Store on Amazon for material from the artist — a new window will open to Amazon.

CJ's Tips: Again, you need to be very exact when asking Alexa for something from iHeartRadio, especially when asking for a radio station. I find saying "Alexa, play 102.7 KIIS-FM Los Angeles" is a bit of a mouthful and can be hard to remember precisely. I prefer to save my radio stations as favorites via the ***Now Playing*** page (as outlined above) and then I find that Alexa is much more likely to understand what I am requesting.

TuneIn

TuneIn is one of the oldest streaming services, founded in 2002. Today, its specialties are Music, News, Sports and Talk, and TuneIn carries local, national and international content. Free and Premium accounts are available:

- **TuneIn Radio:** Free; Ad-supported. Stream 100,000 radio stations covering the range of styles.
- **TuneIn Premium:** $9.99/month; Ad-free. Stream 100,000 radio stations, 600 ad-free music stations, audio books and podcasts; Listen to live coverage of sports events including play-by-play of every NFL, MLB, NBA, and NHL game and more.
- **TuneIn Live on Alexa:** This is a specific deal that was launched in March 2018 for customers who either have an Alexa-enabled device or an Amazon Prime membership. It is effectively an add-on to the Free TuneIn offer, and allows you to access the live coverage of sports events including play-by-play of every NFL, MLB, NBA, and

NHL game and more. The price is currently $2.99/ month ($3.99/ per month for non-prime members that have an Alexa-enabled device). Note that this offer is mainly aimed at owners of Amazon's Echo products, and is not currently fully integrated with some third-party Alexa-enabled devices such as the Sonos One smart speaker.

Linking TuneIn to Alexa: On the *Music, Video & Books* page, select *TuneIn > Link your account > Authorize the link*. You can create an account there by providing the standard information, choosing the membership level you prefer and providing payment information if you choose Premium.

Using TuneIn on Echo: The TuneIn homepage is set up much like the iHeartRadio page with a search box and more specific means of searching.

- Using the Search **Box returns** uncategorized results, unless you type in the exact name of the show or station you want, and that's a disadvantage compared with iHeartRadio. You're given a very long list of results to scroll through, and so I rarely use the function. Obviously, the more specific your search can be, the better results you'll get. You might have to try several searches using differing words or groups of words to find what you're looking for.

- The **Browse** function is more successful, and here TuneIn has the edge on iHeartRadio. Select from:

    — Favorites, which is empty at first but will be populated as you "Favorite" content you like.

    — Local Radio based on where you've given as your location at Settings > Your Echo > Device Location > Edit

    — Trending list of stations and shows currently popular.

    — Music that is divided into 40+ subcategories for decade and musical genre.

    — Talk with 30+ genre subcategories.

    — Sports with 20+ subcategories including popular US, European and world sports, fantasy football and podcasts.

- News with a short list of trending shows from popular sources such as NPR, CNN and BBC, and two expandable subcategories: More Shows and Recent Episodes.
- By Location allows you to find news from continents and regions around the globe.
- By Language, with more than 90 languages offered.
- Podcasts broken into Music, Sports and Talk categories with a range of subcategories in each.

Once you choose a **category > subcategory** from Search or Browse results, a list of shows will be displayed for you to select from. Using the App for this purpose, rather than voice, is the best method.

CJ's Tips: Even if you know what you want, Alexa might have difficulty with your request if you use voice. For example, I said, "Play ESPN Outside the Lines," a popular show. Alexa replied, "Do you want me to add an outside station to your Pandora account?" No, but thanks for asking. The request, "Outside the Lines, ESPN" had Alexa take me to a random ESPN station.

This again demonstrates that the more content available through Alexa, the more specific you'll need to be if using voice, and your request still might not produce the desired results. I use the App, and when I locate a station, show or podcast I want to return to, I "Favorite" it immediately, so I don't have to go searching for it later.

SiriusXM

SiriusXM started by offering satellite radio services and has expanded into online radio where it's a good fit for Alexa and the Echo. There are 70+ music channels, 20+ talk and entertainment channels, 10+ sports channels, 15+ news and issues channels, traffic and weather, Latin, Comedy and more.

Alexa's relationship with SiriusXM is different than it is with iHeartRadio, TuneIn, Spotify and Pandora. SiriusXM is an Alexa Skill, so the platform isn't supported to the same extent as the other services. We'll cover Alexa Skills in detail in a later chapter. It being a Skill might be why the Alexa & SiriusXM connection is more problematic. I suggest you read the information on the SiriusXM page on the Alexa App for an overview and for ratings from users.

The Sirius XM skill is currently rated just 2/5 stars, with many more negative reviews than positive ones. Problems include connecting with and accessing SiriusXM, Alexa not recognizing login information or saying it is wrong, and linking to SiriusXM causing user's Echo devices to stop working entirely. The last issue is especially concerning, though I haven't experienced it. My experience has been that Customer Service at Sirius isn't up to speed about linking Amazon Alexa with SiriusXM at this writing.

Here's something else that might deter you from opening a new SiriusXM account, and it is due to it being a Skill rather than a fully supported service: you can't search or browse SiriusXM from the Alexa App. Using voice is the only current option. You must know the name of what you want to hear or use trial and error to find what you want. I said, "Play football on SiriusXM." The first response was, "What do you want to hear?" When I asked again, the reply was, "I couldn't find a station called 'football' on SiriusXM."

Linking SiriusXM to Alexa: If you want to proceed and have a SiriusXM account:

- Scroll to **SiriusXM** on the *Music, Video & Books* page.
- Select **Enable** and follow the prompts. If you do not have an account and still think it's worth checking out after the caveats given, sign up for an All Access trial or paid account at www.siriusxm.com/amazonalexa
- Sirius will send you a confirmation email.
- Select the link included to set up your password for your SiriusXM account and return to login to begin.
- Then, come back to the Alexa App to select **Enable** and make the connection.

CJ's Tips: Make sure popups are enabled in your browser or system settings. You might have to restart your computer after enabling popups to apply the change. The SiriusXM Skill can also be Enabled on Amazon's site here: www.amzn.to/2xa89GH

Deezer

Deezer is one of the new kids on the block in the US market, but has been around for quite a while in Europe having been created in 2006 in Paris. A cross between Spotify and TuneIn, Deezer is an internet-based music streaming service. The service currently boasts of having 53 million licensed titles in its library, with more than 30,000 stations 14 million active users per month and 6 million paying subscribers as of April 3, 2018.

To use Deezer with Alexa, you will need a Deezer Premium + account which costs $9.99 / month. Then, similar to SiriusXM, the Deezer service on Alexa is provided via the Deezer Skill. So, once you have registered your Deezer Premium + account on the Deezer website, you can then access the Alexa app and activate the Deezer skill (as you would for other Skills). When you do this, you will be prompted to link Alexa to your Deezer Premium account.

At the time of writing, the Deezer skill did not have many positive reviews, and an overall rating of 2.5/5 stars. In my opinion, there are still a number of bugs that need to be ironed out before the Deezer skill on Alexa is a viable contender.

Other Music/ Radio Skills

It is worth noting that there are many other radio stations that are offering specific Skills on the Alexa App allowing you to connect directly to their online stream, from national sports coverage on CBS Sports to local radio stations such as Radio Milwaukee. If you have a favorite radio station it might be worth checking if it has an Alexa skill so you can enable it and enjoy via voice control. Go to **Skills > Search all**

skills on the Alexa App to explore the possibilities.

Setting Default Music Service

I've mentioned Default Music Services a couple of times now so let's discuss them more fully. To find this option on your PC or Mac go to ***Settings > Alexa Preferences > Music & Media > Choose default music services***. On your mobile device, go to ***Menu > Settings > Music*** and then scroll down to ***Account settings*** and select ***Default services***.

Whichever way you get there, you'll see your current options (they may change) include setting a default music library and a default station service.

Once a default has been chosen Alexa will always search that service first when you make a request like "Alexa, play Avril Lavigne". Choose your preferred library and station service based on which sources you use the most, and remember that if you want music from a different source you will have to state that resource in your request, for example, "Alexa, play Avril Lavigne from iHeartRadio".

One quirk to watch out for, I have Amazon Prime Music set as my default library and station service, but if I want to hear something from My Music Library I still have to specify it in my request and be really specific if I want a particular album. Example I asked Alexa "Play Earth, Wind & Fire" and, instead of tracks by the band Earth, Wind & Fire, Alexa started playing a track called "Earth, Wind and Fire" by Miguel from Prime Music. I tried again "Alexa, play Earth, Wind & Fire from My Music Library" and things got better as Alexa played a range of the band's tracks from across My Music Library, but I specifically wanted to hear the Greatest Hits album so I had to try once more, "Alexa, play Earth, Wind & Fire Greatest Hits" before I got exactly what I wanted.

I just relay this fascinating anecdote to demonstrate that setting a default music service is still no substitute for getting into the habit of being as specific as possible with all your requests!

CJ's Tips: Once some music is playing via your Echo you can control it with your voice with some pretty obvious commands like "What's this song", "Pause", "Turn down the volume", "Resume", "Next song", "Repeat this song" etc.

But did you know you can be a bit more adventurous than that? Try asking "Play dinner party music" or "Play music I can dance to" and see what Alexa comes up with!

Setting Up Multi-Room Music

If, like me, you become such a fan of the Echo offerings that you end up having more than one Echo device in your home, you might want to set your devices up to play your music on all your devices throughout your home.

To be clear, this option allows you to hear the same music on two or more Echo devices at once. Also, ideally you'll be using Multi-Room with your Amazon Prime Music or Amazon Music Unlimited account; though you can also use it with streams from Spotify, Pandora, iHeartRadio, SiriusXM, and TuneIn.

Note that Multi-Room is not compatible with Bluetooth, and it will not work for audiobooks.

The setup is the same as creating a Group (discussed further in the Smart Home chapter) – in fact what you'll be doing is setting up a Group just for music. The best way to do this is via the Alexa App on your mobile phone or tablet. Here's what to do:

- Tap on the Devices icon (the little house with the two switches)
- Next tap the + *plus sign* at the top right corner and select **Add Multi-Room Music Speakers > Continue**
- Choose a Group name from the options offered up, or create your own name (for example, Disco), then tap **Next**
- The App will list all devices with speakers that are compatible;

choose the devices that you want to be part of the Group, and click on **Save**

When you're ready to use Multi-Room then all you have to do is say "Play [music selection] on [group name]" or "Play [radio station name]" on Pandora on [group name]".

Note that you can control the music from any Echo device that is part of the group.

Setting Up Stereo Sound Speaker Sets

A different option for music lovers who have more than one Echo device, is to set up the two devices to create a stereo sound system. Note that this is a completely separate function to the Multi-Room where several devices in separate rooms work together; in this instance what you'll be doing is setting up two Echo devices to work together in the SAME room. Here's how to do it:

- Place your devices in the same room, several ft (1 m) apart from each other.
- Go to the App and tap on the **Devices** icon, and then the **+ plus** sign.
- The choose add **Stereo Pair/ Subwoofer > Speaker Sets**
- Then follow the instructions to nominate which Echo device plays sound on the right channel (right side), while the other plays sound on the left channel (left side). Don't forget to tap **Save** when you're done!

Note that you can also add an Echo Sub to your Speaker Set for extra bass, in which case you should place it on the floor near the other speakers and then pair as above.

6: Tell Me a Story — Books

For any voice command you read in this chapter be sure to use your wake word ("Alexa", "Amazon", etc) first.

I must admit that I was not really that keen on the idea audiobooks (though I do love reading), but actually trying out the option on my Echo whilst relaxing one evening has been something of a revelation, and one that I am enjoying exploring more and more. Here's how you too can listen to a vast library of written material using your Echo:

Audible

Audible is an Amazon company, and its offerings have been significantly expanded in the last few years from just audiobooks to podcasts and original content that sounds more like a radio drama than a book because it is written for voice and read by a cast. For example, the X-Files: Cold Cases audiobook is four hours of material adapted from the series and read by David Duchovny, Gillian Anderson and other original and new actors. Audible books are read in the original reader's voice, not Alexa's.

Currently, Audible boasts over 425,000 titles in their English language catalogue ranging across a wide selection of genres, so there is plenty to choose from!

At the time of writing, the standard gold monthly Audible plan includes:

- 30-day Free Audible trial with two Audible Originals and one free audiobook to keep (Audible Originals are exclusive titles that produced in the Audible studios).
- $14.95/month after, with the option to cancel at any time.
- 3 books per month – one standard audiobook and two Audible Originals - and all books chosen during your subscription are yours to keep forever even if you cancel.

- 30% discount off additional audiobooks.
- If you don't like a book, you have an exchange period for trading it for another.
- Free Audible app that allows you to listen on all your devices.
- Whispersync syncing that keeps your place in an audiobook even when you switch devices.

Get Started with Audible: Here's how to get started with your Audible account

- Sign up for a free trial at www.audible.com
- Sign into your Amazon account.
- Choose **Existing Payment Method** (which will be indicated with its last 4 digits) or choose **Add New Payment Method** and complete its details.
- Select **Start Your Membership.**
- Select your free book.
- Cancel within 30 days to avoid being charged, if desired.

Connecting Audible to Alexa: Once your Audible account is established, it's time to integrate Alexa/Echo with Audible:

- In the Alexa App on a PC or Mac: select **Music, Video & Books > Audible**. The audio books that you have chosen/ purchased will be listed ready to enjoy.
- In the Alexa App on a tablet or mobile phone: select the **Play** icon to get to **Music & Books.** The audio book will simply show up in the Audible section on this page, ready to be played.

Listen to Audible on Echo: Here's how to enjoy your Audible audiobooks on your Echo or mobile device:

- From the Audible section, select the title you want to hear using manual controls or voice.
- It will appear on your Alexa App player.
- Say, "Read my Audible book," and Alexa will ask which book, or you

can request it by title.
- For books in progress, say, "Resume my Audible book".
- Use voice for controls like pause, go forward, go back, go to chapter 7, read louder and similar requests.
- You can also say things like, "Stop reading in 15 minutes" and "Set a 15-minute sleep timer" to have Alexa end reading when you want.

Kindle Books

Amazon's Kindle has been an innovative eBook reader since its introduction. Now you can listen to Kindle books with Alexa and Echo — no Kindle needed. However, you should note that Kindle books are read in Alexa's voice.

You can purchase Kindle books individually or choose a Kindle Unlimited. Here's what's currently offered:

- 30-day free Kindle Unlimited trial.
- Kindle Unlimited is $9.99/month and includes unlimited reading of 1 million book titles (new and recent books are not included) and magazines; note that not all of these million books are available to be listened to as audiobooks!
- Kindle books can also be borrowed, lent and rented. To find out more about these features visit www.amzn.to/2ytsbid
- With an Amazon Prime membership, you can select one Kindle First early release book and borrow one additional book free each month.

Getting Started with Kindle Books: Here's how to create a Kindle account

- If you love audiobooks, consider a Kindle Unlimited 30-day trial, and the books can be read on a Kindle reader or Kindle for PC.
- To start a free trial, sign into your Amazon account from the Kindle Unlimited page (amzn.to/2Of9o6h)
- Choose **Existing Payment Method** (which will be indicated with its last 4 digits) or choose **Add New Payment Method** and complete its details, then select **Start Your Membership**.

- Select Start Your Membership.
- Cancel within 30 days to avoid being charged, if desired.
- If you don't want Kindle Unlimited, purchase, barrow, rent or select free Kindle books.
- If you already have a Kindle account, the books Alexa can read will show up in the Kindle section of the Alexa App.

Connecting Kindle Books to Alexa: Once you have some Kindle Books in your Kindle account here's how to integrate Alexa with Kindle:

- **In the Alexa App on a PC or Mac:** select *Music, Video & Books > Kindle Books*. The audio books that you have chosen/purchased will be listed ready to enjoy.
- **In the Alexa App on a tablet or mobile phone:** select the *Play* icon to get to *Music & Books*. Your book will simply show up in the Kindle Books section of on this page, ready to be played.

Listening to Kindle Books on Echo: Here's how to enjoy your Kindle audiobooks on your Echo:

- From the Kindle section, select the title you want to hear or use your voice to request it hands-free.
- The book will appear on your Alexa App Player.
- Use the available manual control options, if desired, including selecting the chapter you want from the Player Queue.
- Use voice for controls like pause, go forward, go back, go to chapter 7, read louder and similar requests.
- You can also say things like, "Stop reading in 15 minutes" and "Set a 15 minute sleep timer" to have Alexa end reading when you want.

CJ's Tips: Selected books in the Audible and Kindle services can be listened to and read at the same time, something Amazon calls "immersion reading." Not only is the text in front of you on your phone or reader, but it is narrated and highlighted too. Amazon says this feature, "sparks an extra connection that boosts engagement, comprehension, and retention, taking you deeper into the book." Most people either love it or can't tolerate it, so if you want to give it a try, learn more at www.audible.com/mt/Immersion.

Now that we've gone through all your audio entertainment options you might want to return to Chapter 3 (page 26) about the ***Now Playing*** page to familiarize yourself with what that page of the Alexa App offers.

7: Write it Down & Get it Done — Lists

For any voice command you read in this chapter, be sure to use your wake word ("Alexa", "Amazon", etc) first.

Alexa offers convenience you'll enjoy when making a Shopping List or To-do List. Both functions are explained in detail below.

To begin, choose the *Lists* tab from the Alexa App menu to see both *Shopping* and *To-do* headings. You will see three options, at the top the words Create Lists (more about that later) and then Shopping and To-Do..

Shopping Lists

Creating and managing a shopping list to take with you on your mobile app or to print is convenient, and you might find the process fun too, as I do. Here are your options:

Manually: Select *Shopping List* from the Lists page, type things you want to buy into the **Add Item** + box at the top of the page, and select the + sign or tap enter.

Voice: For just one item say, "Add walnuts to the shopping list" regardless of the page showing on the Alexa App. Currently, Alexa seems only able to deal with adding one item at a time, so if you have more than one item for your list, then this series of commands works best: first say, "Add items to my shopping list" and Alexa will say "OK, what shall I add?", say, "Walnuts" (without the Wake Word), then Alexa will say, "I've added walnuts to your shopping list, anything else?", say the next item and the back and forth dialogue will continue until, when Alexa next says "...anything else?", you say "No".

Manage your Shopping List: Whether inputted manually or by voice, the items will appear in the list with the most recent item at the top

Voice: Say "What's on my shopping list," and Alexa will read the list out to you. Then say, "Remove brown rice from my shopping list" or "Remove item number 7 from my shopping list" to take it off the list. Saying "Check off" instead of "Remove" works too. When you next check your list on the App, the removed items will be checked off and in the completed section of the list. Note that you will need to access the list manually via your PC or tablet etc., to actually fully delete the item from the list.

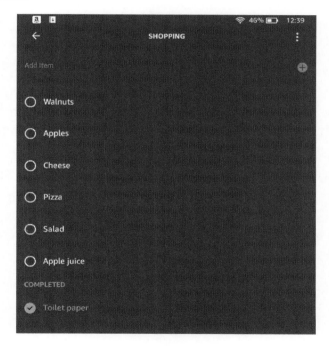

Manually: On your tablet or mobile phone, swipe right to check off an item. It will then move automatically to the completed section of the list. Once in the completed section of the list you can then swipe left to either *Delete* the item off the list or *Restore* the item to the active shopping list. You can also just swipe left straight away to *Delete* the item, or to search for the item on Amazon

On your PC or Mac, check off a purchased or unwanted item by selecting the box to its left, and a ✔ will appear in the box, the item will be ~~crossed out~~, and the delete option will appear on the right. Select the *Delete* option at right, if you want to remove it from the list, or leave it there as a reminder you've bought the item.

Once you go away from the list and return, checked items you haven't deleted will be moved to the *Completed* list where they can be viewed or deleted individually or all at once.

You can also either click on an item, or on the expand symbol (the "v") on the right to see options for searching for the item on Amazon or Bing, and to delete it. Viewing your list on a PC or Mac gives you an added advantage of being able to print items, just click on the *Print* option at the top of the list.

Hear your Shopping List: I say, "Shopping list" to hear it because efficiency appeals to me. If you're more conversational, say "What's on my shopping list," or anything similar. The Alexa voice recognition technology has its voice-recognition "ears" open for "shopping list" almost regardless of what else you say. Try it!

View your Active and Completed Shopping List on the App: On a mobile device or tablet, when you check off an item it instantly appears underneath the "active" list under a heading title Completed; as mentioned before you can then swipe left to either delete or restore the item to your list.

On your PC or Mac, if you go away from your list and then come back to it, the checked items can be accessed via the link that says Completed at the bottom of the active list. You can then uncheck them to have them return to the active list, or delete them if you prefer. Note that you might have to click away from Lists and then go back into the Shopping List to see the items restored to the active list.

To-do Lists

There are just a few differences when creating and managing this list compared to the Shopping List.

Manually: Select *To-do* from the *Lists* page, and type tasks into the *Add Item* + box at the top of the page

Voice: Simply say, "Add call doctor to the to-do list" regardless of the page showing on the Alexa App. Alternatively, say "Add item to my to-do list" and Alexa will say "OK, what can I add for you?" or something similar, then say "Call doctor" (without the Wake Word), and Alexa will confirm that she's done it. Currently, Alexa seems only able to deal with adding one item at a time to the To-do list; and unlike the Shopping List, there is no way of creating a dialogue series for adding more items to the list.

Manage your To-do List: Regardless of how inputted, the items will appear in the list with the most recent item at the top

- Say "Remove repair the faucet from my to-do list" or "Remove item number 5 from my to-do list." Saying "Check off" instead of "Remove" works too.

- Check off a completed task by selecting the box to its left, and a ☐ will appear in the box, the item will be crossed out, and the Delete option will appear on the right

- Click or tap the Delete option at right if you want to remove it from the list, or leave it there as a reminder it is complete

On a mobile device the checked off items appear immediately in the Completed section below. For your PC or Mac, once you go away from the list and return, checked items you haven't deleted will have moved to the Completed list where they can be viewed or deleted individually or all at once..

Hear your To-do List: Unfortunately, saying, "To-do list" won't deliver a reading of the list. Instead, Alexa will ask, "What can I add to your To-do list," or similar. You will have to say, "Read my," "Tell me my," or "What is on my" To-do list.

View your To-do List Active and Completed Tasks on the App: This works the same as for the Shopping Lists described above.

Create Your Own List

A recent update to the App now allows you to create your very own list! Once you land on the Lists page, you can manually add a title in the space where it says Create List, or you can use voice by asking Alexa to create a list! Alexa will ask you what you want to call the list, and then what you want to add... the list you create will work exactly the same way as the To-Do list. I found it to be a handy option for making a guest list for my kid's birthday party.

Deleting or Archiving a List

You might want to delete a list once you've no use for it. The only way to do this is via the App (you cannot use you voice). On your mobile device you can swipe left on the title of your list to either Edit the title, or Archive it. Once archived you can tap on View Archive at the bottom of the page and then in the Archive swipe left on your list again to either Restore the list or Delete it completely.

On your PC or Mac, when you first arrive on the Lists page, you will see the V expand icon on the right and when you click on that you will have the option to archive the list, then view the archive and choose the Restore or Delete options there.

CJ's Tips: When it comes to shopping lists I would say that the major failing is the inability to add more than one item at a time to your list by voice. For example, I have an Echo device perfectly placed in the kitchen and what I want to do is wander around and say "Alexa, add milk, bread, apples, ham and chicken to my shopping list" but that doesn't work. So it works fine when I'm in the kitchen and remember one item I need to add, but if I want to make a long list I'm better off adding it manually via the Alexa App.

As is the case when learning about any Echo device and Alexa talent, some casual practice with creating and managing Shopping and To-do Lists should prove fun and entertaining when a dose of patience is included!

8: Never Forget, Always Be On Time (& More) — Reminders, Alarms & Timers

For any voice command you read in this chapter, be sure to use your wake word ("Alexa", "Amazon", etc) first.

For me, Alexa and the Echo shine as a personal assistant in these categories. While reminders and alarms are similar, let's look at them separately to learn the nuances of each. Note that in the App menu the page label is just **Reminders & Alarms**, but Timers live here too!

Reminders

This is a quick, easy way to avoid forgetting something urgent or important. A friend of mine stops by to see her elderly father several times a week and sets Reminders for him for the following day or two. When she can't get there in person, she calls her dad and does it remotely with her father's phone on Speaker mode.

Reminders can be set and managed manually or with voice, though using voice is easier.

To set and manage Reminders manually:

- Go to the **Reminders & Alarms** page where tabs for **Reminders**, **Alarms** and **Timers** appear.
- Select **Reminders**.
- Select + **Add Reminder**, and a form will appear.
- Fill in the spaces for Remind me to…, Date and Time.
- Select which device you want the reminder to be given on.
- Select **Cancel** or **Save** as appropriate.
- Your reminders appear on the Reminder page in chronological order of when they will occur, not when created.
- Manage any Reminder by selecting it, and you'll be taken to a page where you can Edit it or Mark as Completed.
- If you select **Edit**, you'll be taken to a page where you can edit the Reminder by selecting any of the details or **Delete** the Reminder.
- When editing a Reminder using the Alexa App on a computer, you must click away from the detail you edited, so that it is not highlighted, before the Save option becomes active.

To set and manage Reminders with voice:

- Say "Remind me to call the mechanic at 1pm"
- If you forget to give a time, Alexa will ask for one, and if you say "1," Alexa will ask, "Is that 1:00 in the morning or afternoon?" or similar.
- If you specify time but not date, Alexa will set up the reminder for today, so be sure to give the day, such as "tomorrow" or "Wednesday" for days this week, or give a date such as October 6.
- To cancel a Reminder, use the day and time rather than what you wanted to be reminded of, so say, "Cancel the reminder for 1pm today," and Alexa will ask for clarification if needed.

- **Note:** You cannot currently edit a Reminder using voice.

Completed Reminders: Select the Completed Reminders tab from below the active list to view them. You're not currently able to delete them from this list. They might jog your memory about whether you did something on the list, though being reminded of it is not a guarantee you followed through. For example, a completed reminder on this list doesn't mean a person remembered to take their medication or feed the dog!

Alarms

Alexa gives you a range of alarm options, which we'll get to shortly, but first, here are the basics. Oddly, you currently cannot set an alarm manually, only using your voice, so let's begin there.

Setting and managing Alarms using voice:

- Say, "Set an alarm for 6am"
- Alexa will display the alarm and say, "Alarm set for 6am"
- Say "Cancel the alarm for 6am," and Alexa will say, "6am alarm canceled"
- If an alarm is on and you request another alarm for 7am, Alexa will say, "Second alarm set for 7am," and both alarms will show on the App
- For repeating alarms, say, "Set an alarm for":
  — 6am every day
  — 2pm every Friday
  — 7:30am on weekends
- If you forget to specify AM or PM, Alexa will ask, "Is that 6 in the morning or the evening?" and you can reply without using the wake word as long as the light ring on the Echo is green-blue to indicate Alexa is still listening
- If you don't respond, Alexa will ask again after a few seconds, and if you don't respond to the second query, no alarm will be set.

- All Alarms appear on the main Alarms page with boxes turned On/ Off, indicators shown such as "Every day," with On alarms listed first.

Managing Alarms manually:

- Go to **Reminders & Alarms** and select **Alarms**.
- Use the box next to any alarm to turn it On and Off as desired.
- Select any Alarm to edit it, and a page will display where you can edit the Time, Alarm Sound, choose Repeats options including Never and Delete the alarm if you wish.
- When you've edited the alarm, select **Save Changes** or **Cancel**.

Manage Alarm Volume & Sound manually:

- On your PC or Mac **Manage alarm volume and default sound** from the main Alarms page, and you'll be taken to a page in **Settings** (though here, we discuss just sounds related to Alarms); on the App on your mobile there is a **Settings** link at the bottom of the page for the alarms.
- Change the Alarm, Timer and Notification Volume by dragging the button on the slide bar (Note that this does not change the volume for any other device function, only the volume of the alarm itself).
- Under Custom Sounds, the **Alarm** box will show the Alarm Default Sound, which can be changed by expanding the box using the > shaped caret.
- On your PC or Mac, at the top of the list of **Alarm Default Sounds**, select Celebrity to see an expanded list of celebrity or novelty sounds.

CJ's Tips: If you set up an alarm to go off at the same time every day of the week you won't be able to cancel that alarm for just one day if you don't need it, you have to cancel the whole weekly alarm. Consequently, although it takes me a little longer to set up, I tend to set up an alarm for each day that I need it giving me full control over each alarm.

Also, do check out Celebrity alarm sounds, these change over time and some of them are pretty funny. Failing that, the more traditional alarm sounds are worth exploring to find the one that suits you best; or you can also opt for your favorite song to be the sound that plays for your alarm

Timers

If you're like me, you may well use Timers for a range of purposes, some which overlap with Reminders. These are simple functions, so this can be quick:

- Timers can be set for 1 second to 24 hours.
- Timers **must** be set with voice, so say, "Set a timer for 45 minutes" or "Cancel the timer".
- You can name a timer so try saying, "Set pizza timer for 10 mins"
- Multiple timers can be set to run concurrently.
- Active timers will display and count down on the App.
- Note that timers can only be paused through the App; do this by tapping on them when they're displayed on a device with a touch screen, tablet or mobile phone, or by selecting them on your PC or Mac.
- Select **Manage timer volume** to be taken to the Settings page to adjust Alarm, Timer and Notification Volume.

CJ's Tips: Multiple named timers are so great in the kitchen when you're cooking. Set and name timers for every element of the meal you're making so each timer also becomes a reminder of what's finished cooking or what needs to happen next.

9: Keep it Simple — Routines

For any voice command you read in this chapter, be sure to use your wake word ("Alexa", "Amazon", etc) first.

Routines is one of the newer Alexa features, and one I'd been waiting on for quite some time. The number of things that I can do with Alexa and my Echo devices has grown rapidly since their launch. This also means that the amount of time I would spend "talking" to Alexa has grown too. Every time I wanted Alexa to do something I would have to give her a voice command for each thing and sometimes that could be a drag. I previously mentioned in the Lists chapter that if I want to add items to my shopping list by voice then I have to do it one item at a time... "Alexa add oranges to my list," "Alexa, add apples to my list," "Alexa, add bananas to my list" etc, etc. If I have 20 items to add to my shopping list at once this can get really old, really quickly.

Well, although the **Routines** feature doesn't yet solve that little conundrum it is a step in the right direction and shows me that Amazon is now addressing the question of how to activate several functions at once with just one voice command.

So here's the idea of Routines. We all have daily habits and routines, from getting up and dressed in the morning to coming home from work and cooking dinner in the evening. Now you can group certain Alexa features together to compliment those daily routines and have them all start/activate at the same time, either with one single voice command or at a set time each day.

For example, if you wake up at the same time each day you can program a morning routine for Alexa to turn up your thermostat, turn on your Smart Home kettle, give you the weather forecast and catch you up on news headlines via your Flash Briefing. Either set this routine to start at 6.30 am every morning or start the routine by saying something like "Alexa, start the day" or "Alexa, good morning" or whatever you like!

So that's the good news about Routines. The slightly less good news is that, at the time of writing, the actions that you can program as part of a routine are limited to playing music, Smart Home devices, your calendar, your Flash Briefing, your traffic information and a weather forecast. Personally, while I do use Alexa's news, traffic and weather features, they are not so important to me that I need to schedule them for a set time each day. Bunching the operation of several Smart Home devices into one request is certainly welcome but I already have the ability to do that via Smart Home Groups (discussed later in this guide).

So Routines is a great idea and one that I think will become very useful in the future as more options are added. I'm hoping that it won't be too long before Reminders are added to the list, as that would be a very useful thing to hear first thing in the day, along with the weather report and the radio playing.

Another thing to note is that at the time of writing, I could only access the **Routines** page of the App via my tablet and mobile – and not via my PC.

For now, let's look at how to setup a routine. Go into your Alexa App on a mobile device and tap the **Menu icon** (top left) and then **Routines**. You will see that Alexa has some suggested Featured routines. For now, ignore this and tap the big + **Create Routine** in the middle the page, of the + **sign** at the top right of the page.

Now you will have the options to **When this happens** and **Add action**. Tap the + sign next to **When this happens** and chose and option – the obvious being either to create a voice command for your routine (example, "Alexa, good morning") or schedule a set time for your routine. When scheduling a time, you not only set the time of day but can choose to have the routine happen on a specific day, every day, week days or weekends. You can also link your routine to a certain Smart Home device activating, or the pressing of an Echo Button, or for when you dismiss a particular alarm.

Follow the prompts to save what cue or condition is used to activate the routine. Don't forget to press Save when you're done.

Now tap the **+ *sign*** next to **Add action**. You will now be able to choose from the options I mentioned earlier, **Music, News, Smart Home, Traffic, Weather** or **Alexa Says** (this last option is quite fun – offering a number of stock phrases that Alexa can use to reply to your Routine opener). There is also an option to set the volume for your Routine feature marked ***Audio Control***. Setting up some of these features are discussed in the relevant chapters of this guide. For now, all you need to know is that you can tap **Add** to include one or all of these features to your new Routine and then tap ***Create***.

You may also be prompted to choose which Alexa enabled device plays the routine, if you have more than one Echo device for example. Again, don't forget the tap on ***Save*** when you've got everything set up how you want.

Your new routine will now appear in your list of routines and you can tap again on any routine to edit it, temporarily toggle the routine on or off, and delete the routine if you wish. I suggest you read on now and return to this chapter when you have read the Smart Home chapter (page 81) and the Things to Try chapter (page 88) where I discuss linking Smart Home devices to Alexa and setting up news, traffic and weather briefings.

CJ's Tips: In the upcoming Smart Home chapter I discuss how to put several Smart Home devices into Groups for convenience. One thing that Routines can do, that Smart Home Groups can't, is to activate a number of devices at a set time of the day (or night). This is an aspect of Routines that is already very useful for me. I like certain lights in my home to come on and switch off at certain times of the day, especially when I am away from home to give the impression that the house is still occupied. Scheduling these lighting states for set times and on set days is now a breeze with Alexa Routines.

10: Let's See What She Can Do — Alexa Skills

For any voice command you read in this chapter, be sure to use your wake word ("Alexa", "Amazon", etc) first.

Alexa Skills are like "apps" for your Echo. In the same way that you download apps to your tablet or phone for added functionality, you can add (enable) Skills for Alexa that allow you to do more and get more out of your device. Adding and using these skills is super easy and there are currently more than 70,000 Skills to choose from, and the number grows daily! Skills are loosely grouped into 23 categories currently. Some are good, many are bad, there are sure to be a few of interest for every Alexa user. For an overview of Alexa Skills visit: www.amzn.www.amzn.to/2zo5vBM

Get Familiar with Skills

Start your Skills adventure by getting to know what's available. There are three ways to browse Skills, and all begin on the Skills page on the Alexa App Menu (Skills & Games on the mobile App version).

This is one area of the Alexa App that I think is easier to deal with on your PC or Mac, since there are so many Skills to consider having a larger screen to view them on makes it easier. So, ideally on your PC or Mac, you'll see that the default setting on the Skills landing page is to show All Skills. Near the top right is a tab called Your Skills, which will be populated as you enable Skills for your use. Once you are adding Skills, you'll likely toggle back and forth between the All Skills and Your Skills lists quite a bit. OK, let's explore starting on the Skills page in the Alexa App.

If you access the App via your mobile, the Skills & Games landing page defaults to a page called Discover skills that has various rotating and scrollable tabs that link to information about certain Skills or skills categories. Along the top of the landing page you'll also see where to go to search for skills via Categories; and a link to Your Skills, which will be populated once you've enabled some.

So here's how to find Skills that will enhance your Alexa experience.

Option 1: Scan the rows of Skills on the All Skills or Discover Skills landing page.

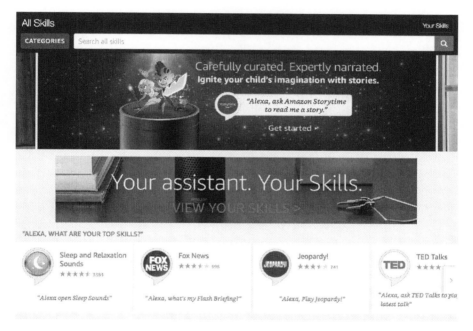

Option 2: Explore the **Categories tab** where Skills are organized in over 20 categories including:

- Newest Arrivals
- Business & Finance
- Communication
- Education & Reference
- Games, Trivia & Accessories
- Kids

- Lifestyle
- Movies & TV
- Music & Audio
- News
- Productivity
- Shopping
- Smart Home
- Social
- Sports
- Travel & Transportation
- Weather

Select any category to browse the Skills in it. There will likely be several Skills and you can organize your browsing by choosing the "Sort by" options to the right of the results (Relevance, Customer Rating, Release Date or Featured).

CJ's Tips: These rows are followed by a list of all the Skills in the category. It can be a long list! Currently, there are nearly 900 Skills in the Health & Fitness category, so rather than scrolling through the options, I used the search box to narrow the list to skills related to a personal interest such as biking. Sixteen results appeared. If there were still too many results, I would use the "Sort by" options to narrow the search further.

Option 3: Use the Search box to find a Skill for something very specific (e.g., ecobee, weather Canada, Corvette) or to see if there is one related to one of your interests. For a random sampling, I typed in:

- "Chinese" and received 72 search results related to learning the language or one of the dialects, cuisine, the Chinese calendar and zodiac, etc.
- "Flowers" — 40 results about ordering flowers, facts and trivia, state

flowers, etc.

- "Michigan" — 46 results about trivia, history, several state universities, winning lottery numbers, snow reports for skiing, information from news outlets, etc.

Again, the results can be narrowed using the **Sort By** options at the top right of the results list or by adding a word or two to your search query — "Chinese language", or "University of Michigan", for example. For something like "news" with more than 2,500 results, you'll have to search a specific news organization, news about a topic, city, event or similarly narrow term.

Enable Alexa Skills and Use Them

When you're ready to put a Skill into action, the first step is to enable it, which is like turning it on for use with your Alexa account and Echo.

It's important to note that some of the Skills require creating and/or linking to an account or subscription separate from Amazon, so additional steps might be required. Recently, I enabled the Allrecipes Skill and discovered I had to open an account on the Allrecipes site and link to it to get the most from the Skill.

Here's how to enable and use an Alexa Skill:

- Use one of the options above to locate a skill you want to enable, and select the skill
- Select **Enable Skill**, and after a few seconds, the box will switch to **Disable Skill**, which means it's been enabled and will appear in the Your Skills list.
- Note that you might be required to link the Skill to your Amazon account, in which case follow the prompts to do so – this will usually require you to log into your Amazon account.
- Alternately, say, "Enable the Rainforest Sounds Skill" (for example).
- You can also browse and Enable Skills on Amazon.com
- To use any Enabled Skill, include the name of the Skill when making

a request such as "Alexa, play rainforest sounds".

- **Note:** Skills in use do not appear on the *Now Playing* page or Player, so to control them, say "Stop," "Pause," "Resume" or similar.

Get the Most from the Skills You Enable

The more familiar you are with the Skill's capabilities, the more it will benefit you. Here are suggestions for optimizing a Skill's usefulness:

- If another action is offered along with Enabling the Skill, such as Manage in News Briefing (common for Skills from news organizations, and covered later), select the option to learn about it and decide whether to use the feature.
- Read the *About the Skill* section to acquaint yourself with its capabilities.
- Review the *Try Saying* suggestions to get the Skill to do what you want and additional suggestions in the About information.
- Learn the *Invocation Name* for the Skill (found under Skill Details) to be sure Alexa will put it to work when you want to use it.
- Browse the *Customers Have Also Enabled* section for similar or complementary ideas.
- Read reviews, and, if interested, add a review after you've used the Skill for a few weeks.
- Select *Having Trouble with This Skill* to get help using it from the source of the Skill.

Disable an Alexa Skill

Over time, the Your Skills section can become crowded with Skills you don't use. They're easy to remove.

- Select the Skill from Your Skills list.
- Select Disable Skill, and the box will flip to Enable Skill to indicate it isn't enabled and the Skill will disappear from your list.

CJ's Tips: Whilst researching this book I've explored more than 800

Skills and Enabled more than 500, but my Skills list currently has 48 Skills; I use some daily and others only occasionally.

The Skills I rely on the most are those that save me time, are easy to use and/or offer genuine value. If they're a hassle, a time-waster or not the easiest or best way to do things, I either disable them immediately or as soon as the novelty wears off or when I occasionally go through the list to get rid of unused Skills. Some have features that are easy to use and features that are a hassle. For example, when I have a specific recipe in mind, I bring it up on the Allrecipes Skill and it's a delight to use. However, browsing recipes to get meal ideas is tedious using the Skill, so I go to the Allrecipes.com site on my computer or phone.

You'll soon see that there are sometimes several Skills for some purposes. Searching Skills for "find my phone" yielded seven results, "restaurant finder" showed ten options and "control lights" produced a list of 48 smart home Skills. In the case of light control, some of the Skills require a specific brand of home automation equipment. Look for your brand's logo in the results list. If you don't find it, explore some of the top-rated Skills by reading the About information, Requirements for its use and Reviews. Reading the information is a good approach when choosing any Skill from a results list.

When I'm undecided about which of two or three Skills to choose, I enable both or all and use them alternately. I often use one right after another, and do that every day for a week or so. This is a better way to determine which Skill is right for me than to try one for a week and then try another. I like an immediate, one-after-the-other comparison.

The point I'm delicately trying to make is that there are many Skills that...just aren't very good. It's worth putting in some time to find the gems that you'll use over and over, but be prepared to wade through quite a lot of rubbish!

Alexa Skills Blueprint

This is a very recent development at Amazon, and one that I'm looking forward to exploring further. Basically, the idea of Skills Blueprint is that you can use pre-loaded, customizable templates provided by Amazon to create your own unique Skill.

For example, you could set up a care guide for your pet sitter – with feeding instructions and such, or create a trivia game skill about your family history to play at family gatherings; or you could build your own bedtime story – including sound effects – where you and/or your kids are the heroes!

If this is something that appeals to you, go straight to the Amazon Skill Blueprint site at blueprints.amazon.com; click on the template that interests you and follow the instructions to create your very own Skill which is then be associated with the Alexa-enabled devices that are linked to your Amazon account. Simple!

11: 21st Century Living — Smart Home

For any voice command you read in this chapter, be sure to use your wake word ("Alexa", "Amazon", etc) first.

If you're already integrating smart technology into your home, then learning to control it with Alexa's help will be one of the easier parts of the curve. There's a four-step process that is straightforward and usually successful. If you're new to smart home devices then I recommend you visit the Amazon Smart Home page (amzn.to/2wHRFJW) to familiarize yourself with the possible options.

With the Echo, you'll be treating your smart home tasks much like Skills and used as such. You'll be well prepared for the information here if you've read the previous chapter and enabled a few Skills. But before we go into further detail, Amazon suggests we cover a few common-sense security tips for using smart home devices with Alexa, so let's do that first:

- Follow the device's instructions for safe, recommended uses.
- Confirm that requests have been carried out — especially tasks related to your home's safety and security (security system, exterior lighting, door locks, garage door, HVAC, appliances, and similar).
- Turn off the microphones on Echo and other Alexa devices if you do not want Alexa to respond to voice commands when safety and security cannot be ensured (such as when adults are away from home).
- Remember that once a device is connected, anyone can use Alexa/Echo to control it, so make sure those in your household and guests understand safe operation of smart home devices.

Okay, let's get rolling.

CJ's Tips: With regard to setting up smart home devices and Alexa, I find it much easier at the start to do the setup via the Alexa App on my PC – the App interface is just a lot clearer to follow. After setting things up, I am then able to use the App on my tablet to help manage my smart home devices. Note that on the App via a tablet/ mobile phone you need to tap on the icon depicting a house with two switches inside which takes you to a landing page for all your smart home Devices. You cannot get to this page via the Menu on the tablet/ mobile version of the Alexa App.

For the rest of this chapter, *please note that the instructions given are for when you view the Alexa App via your PC or Mac.*

Step 1: Prepare your Smart Home Device

This step ensures your device is compatible and optimized for use with Alexa:

- Verify that your smart home device is compatible with Amazon Alexa, which can be done by searching for it on the Skills page of the Alexa App or Amazon.com, checking the packaging, contacting the manufacturer or viewing its website and looking for the device on the Alexa smart home shopping page (www.amzn.to/2wITbeJ) that includes all compatible devices.
- Download the manufacturer's app for the smart home device to any mobile device (preferably the same one that you're using to manage the Alexa App).
- Use the manufacturer's app to set up the smart home device/ equipment on the same Wi-Fi network your Echo is on.
- Download and install the latest software updates for the smart home device.

Step 2: Enable the Device's Skill

If your smart home device is compatible with Alexa, it will have its own Skill.

- Locate the **Skill** by going to the main Skills page on the Alexa app, clicking **Categories** then **Smart Home** and using the search box to find the relevant Skill. Alternately, Skills can be searched on Amazon.com (www.amzn.to/2ONKNni).

- Select the Skill.

- Read the **About** information and the **Skill Details** on the Skill page that opens to familiarize yourself with its capabilities.

- Select the **Enable** button, and when it changes to **Disable**, you'll know the Skill is enabled.

- When prompted, sign into the smart home device account to link it to your Alexa account.

- Follow any prompts given to complete setting up the smart home device on Alexa.

- **Note:** A list of your Enabled smart home devices will populate on the **Smart Home** page in your Alexa App. Select any of the Skills to Disable it or to learn more about it.

Step 3: Ask Alexa to Discover the Device

This can be done two ways:

- Say, "Discover devices".
- Go to the **Smart Home** page on the App, select **Device**s, and then select the **Discover** tab.

Step 4: Set Up Groups and Scenes for Your Smart Home Devices

As you connect smart home devices to Alexa, you can then set them up to work together by putting them into groups or scenes.

Basically, when your smart home devices are put together in a Group the command you can give them via Alexa is simply to turn those devices on or off simultaneously, so for example turning on three specific light bulbs in the bedroom at the same time.

However, the Scenes option allows you to predetermine the state of those devices, for example if you want the light bulbs to be on low and the thermostat to go down to a lower temperature at bedtime (in this case the Scene would be named Bedtime and the command to Alexa would be "Alexa, turn on Bedtime").

The other big difference is that you set up a Group within the Alexa App; whereas to set up Scenes you need to first configure those scenes in the smart home device's app, and then those specific device scenes will be listed in the **Scenes** page on the Alexa App, where you can then control those scenes via Alexa.

Effectively the **Scenes** page simply shows you the different scenes you have already set up and then allows you to manage them with Alexa voice commands.

How to Set Up Groups:

Chose the **Groups** tab on the main **Smart Home** page:

- Choose **Create Group** to connect two or more devices that will work together with a single command.

- Follow any additional prompts to complete formation of the Group including naming the group and selecting devices for the group.
- Try the Group by making a request of Alexa appropriate to the Group's functionality ("Turn on the lights," or "Lock the doors," for example).
- To edit the Group, select it and choose the name to change it or select/deselect devices you want in the group.
- To delete a group, select **Delete this group**.

CJ's Tips: Alexa uses voice recognition to understand what you say and connect it with digital content such as the name of a group you type into the App. For this reason, give groups names that won't confuse Alexa. To be on the safe side, I recommend sticking with the Common Group names that Alexa offers.

How to Manage Scenes:

First configure specific Scenes within the smart home device's app. For example, if you want to create a scene with the Philips Hue smart light, then you need to go into the Hue app to set up that scene.

Then choose the **Scenes** tab on the main **Smart Home** page:

- All Scenes that you have set up on your smart home devices should appear, meaning that you can use voice command to activate those Scenes, using the name you've set for that Scene ("turn on Bedtime" for example).
- If you don't want Alexa to control a certain Scene, then click on **Forget** to delete the Scenes from the Alexa App.

Troubleshooting Alexa and Smart Home Device Set Up

If Alexa doesn't discover your smart home device, try these tips in this order. They will troubleshoot and solve most issues.

Device Issues: Double-check that:

- Your smart home device is compatible with Alexa.

- That you have Enabled the device's Skill.
- That you have downloaded the device's app.
- That you have downloaded the latest software updates for the device.
- For a Philips Hue Bridge, be sure to press the button on the Bridge before trying to discover devices.

Note: Additional details for these steps are found in the four steps above.

- Restart your Echo, and restart your smart home device.
- Disable and Enable the smart home device.
- In the Alexa App, choose the Forget option for the smart home device to unlink it from Alexa, and then try to reconnect it.

Wi-Fi Issues: The Echo and your smart home devices must be on the same Wi-Fi network. Personal networks are best; school and work networks you don't control might not allow unrecognized devices to connect.

Check or change the Wi-Fi network for Echo by:

- Choosing Settings for the Echo.
- Observing what network the Echo is on beneath the Wi-Fi option.
- Changing the Wi-Fi network your Echo is on, if necessary.
- Using the Add a Network option to add a network for your Echo and smart home devices.

Note: Some smart home devices can only connect to the 2.4 GHz Wi-Fi band. If you have a dual-band network, make sure it is set to the 2.4 GHz band.

Change confusing group names: Remember that Alexa won't comprehend words with numbers or symbols in them like kitch3n lights or #kitchen lights@, so make changes to group names, if necessary.

Discover your smart home devices again: If you've made any changes discussed here, including the Wi-Fi network or band or a group name say, "Discover devices." Alexa will let you know if they are found or will say "I didn't find any devices." If this occurs, contact the smart home device manufacturer for further assistance, since it is up to the device manufacturer, not Amazon, to ensure that their devices are compatible with Alexa devices.

How to Disconnect Smart Home Devices from Alexa

A list of your smart home Devices will populate the **Smart Home** main page. Select any of the Devices and then select **Forget it** to disconnect it from Alexa.

12: Your Highly Capable Personal Assistant — Things to Try

For any voice command you read in this chapter, be sure to use your wake word ("Alexa", "Amazon", etc) first.

This is a large section of the Alexa App and could be a book of its own!

The good news is that many of the **Things to Try** have been discussed in other sections, and the ones I don't talk about anywhere are very straightforward and easy to use with the few words of explanation given in the Alexa App. Our purpose in this chapter is to shed light on Things to Try that haven't been mentioned but require some explanation.

To get started, select the **Things to Try** from the menu in the Alexa App to explore them for yourself or to follow along with this discussion. The interface of the Alexa App for this page is quite different when viewed on a tablet or mobile phone compared to a PC or Mac. On a PC or Mac, the Things to Try section is just a list of items, whilst on a tablet or mobile phone the list is presented with both words and icons for each item.

I have also noticed that there appear to be many more **Things to Try** items listed on the tablets and mobile phones version, than on the PC or Mac version of this page – an indication, perhaps, that more and more users are primarily accessing the Alexa App on their mobile devices.

Either way, when you select any of the items on the list, you will usually be taken to the **Help & Feedback** section for an explanation. The list serves as an easy place to find topics you want more information about

I have also noticed that there appear to be many more Things to Try items listed on the tablets and mobile phones version, than on the PC or

Mac version of this page – an indication, perhaps, that more and more users are primarily accessing the Alexa App on their mobile devices.

Either way, when you select any of the items on the list, you will usually be taken to the **Help & Feedback** section for an explanation. The list serves as an easy place to find topics you want more information about.

Here are some of the useful and interesting **Things to Try** that haven't been detailed already in the book:

What's new?

I like to keep up with Alexa's capabilities, so I check this section weekly, at least. There's always something new. It's a mixed and disorganized bag ranging from simple, fun stuff ("Drum roll") to interesting information (the ability to give publication dates and authors for popular books) to new Skills ("Give me a quote") to new Alexa-enabled devices. It's a potpourri of fresh possibilities with Alexa and Echo.

Ask Questions

If you can think it, the question might be worth asking. I ask these types of questions quite often:

- What time it is here or elsewhere?
- Conversions of weight, measurements temperature, money.
- Spellings/synonyms/definitions.
- People and place facts.
- Math.

Check and Manage Your Calendar

Alexa can link to calendars from Google, Apple and Microsoft.

- Go to **Settings**, and scroll down to the **Accounts** section to **Calendar**.
- Choose the calendar you want to connect, and select Link _____ account.
- Link another calendar, if desired.
- Say, "What's on my calendar?" "What's my next appointment?" or similar.
- Say, "Add a 9:30am meeting to my calendar".

Discover new Alexa Skills

See the chapter dedicated to Alexa Skills (page 71) dedicated to Alexa Skills, but if you want to hear about the latest releases:

- Say, "What are popular Skills?"
- Say, "What new Skills do you have?"
- Browse the Skills store mentioned earlier where every Skill can be

found and Enabled (www.amzn.to/2xOQ9pi).

Find Local Businesses and Restaurants & Reviews

If you haven't set your location yet, this is a good time to do that, so Alexa knows where to search. On your PC or Mac go to **Settings**, select your device. In the **General** section of the Alexa App, select **Device location** and **Edit**, if the location is incorrect. Add as little as your zip code or as much as your full address. The more information you provide, the more accurate Alexa can be in telling you how far you are from businesses and restaurants.

Note that this task can be done on the App via a mobile/ tablet by going to **Menu > Settings > Device settings**. Then choose your device and scroll down to **Device location**. Select that option and you can enter or change your address there. Don't forget to tap Save when you're done.

Now you can ask Alexa to help you find local businessesHere are things to try:

- "Find a Mexican restaurant".
- "Is the post office open?"
- "What clothing stores are nearby?"
- "What supermarkets are near me?"
- "What is the phone number for B&V Asian Market?"

Your search doesn't have to be local either. When planning a trip, you can ask Alexa to find any of these things in another town.

- "Is there a shoe store in Grand Rapids?"
- "Pharmacy in Plano, Texas".
- "Is there Crossfit in Raleigh, North Carolina?"

As well as Alexa reporting the result of your search to you, all searches produce Cards, so you'll have the results right in front of you on the Alexa App with detailed information including, where applicable Yelp ratings.

Search results might be varied, especially when there are only a few good options. When using Alexa in a rural area, I requested a "sporting goods store," and got one accurate result and four others, ranging from a used firearms dealer to a marina. See how you do with your searches!

CJ's Tips: Alexa is far from perfect yet, but I'm often impressed with the information I can get from her. If I need to call a local shop or business I always ask Alexa for the number first before reaching for the phone book or doing a search online; more often than not, I get the info I'm looking for.

Find Traffic Information

Alexa can give you an estimated time for your commute and the fastest route.

- In **Settings** on the Alexa App, scroll down and select **Traffic**.
- Add or change your current address, your "From,".
- Add a destination, your "To".
- Stops along the way can be added too.
- Later, select **Change address** to edit any of the locations.

Once set, you can then ask Alexa, "What's my traffic?". While I use this for the drive I make most often, it can be convenient for determining the time for other driving routes by changing the "From," "To" or both.

Get Weather Forecasts

Once Alexa knows your location in **Settings > Your Device > Location,** you can hear local weather information on Echo and view the forecast on the Card produced. Ask about weather in other cities around the globe too. Common questions Alexa can answer include:

- "What's the weather?"
- "Show me the forecast".

- "Will it rain tomorrow?"
- "How warm will it be today?"

There are many ways to answer the question, but Alexa replies to all of them with the same information — verbally sharing current and expected weather conditions for the next 24 hours and showing an extended forecast within the App.

Go to the Movies

This section uses your location to access local movie schedules and can give you information for movie schedules in cities you'll be visiting. I get the best results with questions like these:

- "What movies are playing?"
- "What movies are playing at [name of theater complex]?"
- "When will Indiana Jones play tomorrow?"
- "Show me the trailer for the movie Dunkirk".
- "Tell me about the Aquaman movie".

Hear the News

Alexa offers something called a **Flash Briefing**. Many news and entertainment organizations including NPR, NBC, FOX, BBC and ESPN make brief overviews of the news, such as you might hear at the top or bottom of the hour, and those summaries can be added to your Flash Briefing.

To select which summaries you want to hear in your Briefing, and to start and manage it:

- Go to **Settings > Alexa Preference > Flash Briefing** to toggle On or Off the default options, if there are any.
- If not, go to Skills and search your favorite news, sports, weather and entertainment stations (there are new media coming onboard every week), and enable the Skill.
- If it offers a Briefing summary, that will show in your list of Flash

Briefing choices in Settings.

- Toggle On or Off the briefings, and you can make changes when desired.
- Ask "What's the news," "Give me my flash briefing," "Flash briefing," "What's in the news?" or something similar, and you'll hear the entire briefing, one organization at a time.
- Note that the briefing does not show on the Now Playing page or the Player.

CJ's Tips: Once you've toggled the briefings ON, a blue Edit Order options will appear at the top of the page. Click on this and then tap/click and drag each of your news sources into your preferred order of importance. I find that there are a couple of news sources I value above the others, so I definitely want to hear them first. Don't forget to tap/click Done when you're finished to preserve the order you've chosen.

Get Your Sports Updates

Give Alexa a team name and say, "score," and you'll hear the score of the most recent game. If you want to hear the latest scores and the next game for all the teams you follow, go to **Settings > Sports** to search and select those teams. Then, you'll get the answer when you ask questions like:

- "Sports scores".
- "When do the Seattle Mariners play next?"
- "NBA scores".
- "Score for Real Madrid".
- "How many touchdowns did Odell Beckham junior score?"
- "How many homeruns does Aaron Judge have?"

Many major football, baseball, basketball and soccer leagues are currently supported on the Alexa App, including:

North American Leagues:

- MLB — Major League Baseball
- MLS — Major League Soccer
- NBA — National Basketball Association
- NCAA — National Collegiate Athletic Association
- NFL — National Football League
- NHL — National Hockey League
- WNBA — Women's National Basketball Association

European Leagues:

- English Premier League
- FA Cup — Football Association Challenge Cup
- German Bundesliga
- UEFA Champions League

CJ's Tips: Above is a list of all the teams officially supported via Alexa, but I've found that I can get information for many other teams just by searching for them in the search box at **Settings > Sports Update** within the Alexa App.

Shop Amazon (for Prime Members)

Did you hear about the 6-year old girl in Texas who ordered a dollhouse with Alexa or Jimmy Kimmel's prank of telling Alexa to order pool noodles during his show broadcast into millions of US homes?

These are cautionary tales about shopping via Alexa, but with precautions, you'll be immune from such surprises.

To shop on Amazon, you'll need:

- A Prime membership (a 30-day free trial is available).
- A US shipping address.

- A payment method stored with Amazon in your account's 1-Click settings.
- Voice Purchasing enabled in the Alexa App at **Settings > Alexa Account > Voice Purchasing**.
- (Optional) A 4-digit code in **Settings > Alexa Account > Voice Purchasing** that you'll have to give as a safeguard against unauthorized shopping, accidental purchases and not-so-funny pranks.
- Use **Manage 1-Click Settings** in this section to be sure you have a payment method connected to 1-Click.
- An Echo device and the Alexa App.

Once you're set up, you can browse via the Alexa App **Homepage** and order Prime-eligible items with Alexa. Here's how my shopping with Alexa usually goes when I'm not sure which product I want:

- I say, "Shop coffee makers," and then browse them in the App.
- I select Details of any product to learn more including color options, sizes or to **Add to Cart**.
- When I make the decision to buy, I select "Buy This," and the order is placed.
- If I get buyer's regret immediately, I say, "Cancel my order," and it's cancelled.
- Canceling orders after a longer time must be done online
- When the order ships, Alexa gives me a voice notification that also shows up in **Settings > Notifications**.

When I know what I want, the process is much shorter:

- "Order _____," and Alexa will ask you to confirm the order or will show several options on the App page.
- "Reorder _____," and Alexa will ask for confirmation.
- "Add _____ to my cart," is an option too, and you can then go online to review your cart and place the order.

You can also buy digital music with Alexa if you have a US billing address and payment method stored with Amazon that is issued by a US bank. Amazon.com gift cards can be used too.

Physical goods ordered can be tracked with a simple request, "Track my order."

Calling and Messaging

This Echo feature has tremendous potential and will become more useful as the universe of Echo users expands. If you know three or more people with Echo devices, this section might be worth reading now. Otherwise, it's here when Echo catches on more broadly among your friends, family members and business contacts. You can communicate with other Alexa users if you have both downloaded the Alexa App (note that you don't necessarily need to have a specific Amazon Echo device to download the Alexa App onto your phone), and inputted the contact's information correctly.

Note that at the time of writing, I could only access this section of the ***Things to Try*** through my tablet and mobile phone, so I suggest that this is how you do it too.

Setting up Calling & Messaging

- Ensure you have the Alexa App on your tablet of mobile phone.
- Download the latest Alexa App for iOS 9.0 or higher here: https://apple.co/2xhWycQ
- Or for Android 5.0 or higher here: ***https://bit.ly/2fBWcUi***
- Select the ***Conversations*** icon (a speech bubble) at the bottom of the page, and follow the instructions to sign up for Alexa Calling & Messaging and to verify your mobile number.
- Import your contacts, and those who have signed up for Calling & Messaging will appear in your ***Contacts*** list.
- To add or edit contacts for this service, update your phone's local Address book and then open the Alexa App.
- Give your contacts names you'll remember, ask Alexa to call one of

them, and Alexa will dial the number associated with that name.

- Calling & Messaging can be used on Echo, Echo Show, Echo Dot, Echo Auto and Fire Tablet with Alexa, or between non-Echo users that have the Alexa App on their phone.
- Note that some Echo devices with a screen offer a video calling option.
- If you ever wish to deregister from Calling & Messaging, contact Amazon customer service at 1.877.375.9365.

Using Calling and Messaging for Calls

- Tell your Alexa-using friends and family about Calling & Messaging, and encourage them to get set up.
- To make a call from your device say, "Call Aunt Ellen," and Alexa will begin the call.
- To make a call from the Alexa App, select the Conversations icon, select a contact and the Phone icon for an audio-only call (or the Camera icon for a video call, if applicable.)
- When someone is calling you, Alexa will let you know who is calling.
- To answer a call, say, "Answer" or select the Answer button on the devices screen, if applicable.
- To ignore the call, say, "Ignore the call."
- To end a call, say, "Hang up" or tap your touchscreen and select the end-call button.
- There's a video about calling with Alexa here: https://amzn.to/2AOKbbf

Using Calling & Messaging for Messages

- Messages are recorded and played back to the recipient rather than transcribed and read by Alexa.
- To send a message from the Echo, say, "Send a message to Sara Smith," and Alexa will prompt you for the message and send your message when you complete it.
- To send a message from the Alexa App, select the Conversations

icon, choose a contact and then select the Keyboard icon; type a message before hitting the Send button.

- To reply to a message from the App, select the conversation from those shown, and press the Microphone icon while speaking your message, releasing it when done to send the message or sliding left to cancel the message.
- To hear messages on the Echo, say, "Play my messages".
- When more than one phone number is synced with Alexa, you can name the numbers and say, "Play messages for _____" to hear only your messages.
- To review messages on the App, choose the Notifications icon or go to Conversations to view those with a New Notification icon, and then choose between hearing the messages and reading a transcription of them..

There's a video about calling with Alexa here: www.amzn.to/2wSzRfk

Drop in

- Drop in is typically set up among two or more Echo devices in a single household, between family members and BFFs.
- Drop in means that people in separate parts of the home or in different homes can communicate immediately without making a formal call
- Drop in must be enabled in the mobile Alexa App for all devices that will use the capability.
- In the **Conversations** section of the mobile app, select a contact and turn On the Drop in button if they have drop-in privileges with you.
- To drop in on one of your own devices, say something like, "Drop in on my living room Echo".
- The names of the devices in your home can be changed in the Settings on the Alexa App.
- Friends and family with Alexa can also drop in on each other, but

you must both first enable Drop In in your Contacts list. Go to the **Contacts** section of the App, select a contact and turn On the Drop-in button if you want them to have drop-in privileges with you. Note that, on the Communications page of the Alexa App on a mobile device you can also see your **Contacts** by tapping the icon of a person on the top right corner of the page.

There's a video with complete Drop in details here: www.amzn.to/2vJs26V

Announcements

A relatively new addition to the options for Calling & Messaging, the Announcements option works like a one-way intercom where you can basically announce a short message, such as "Dinner's ready!" or "I'm one my way home" to other Alexa enabled devices on your account. For example, from your Echo Show that's in your kitchen to the Echo Dot that's in your kid's room and the Echo that's in the home office. Or from your Alexa App on your mobile phone back to your Echo devices at home.

This is basically similar to Drop In, but just for short voice-only messages. Or you could think of it as a voice version of a text message! Here's how to do it:

- You can use your voice to make an announcement, say "Announce (Dinner's ready!)" and your other enabled Alexa devices will chime and play the announcement.

- Alternatively, manually on the App you can select the Communication icon, then the Announce icon (a loudspeaker) and then type in your message or opt for the microphone to record it; finish by selecting the arrow icon to send.

- Note that mobile devices with the Alexa App (ie your mobile phone) and Echo Auto can send but not receive announcements.

While I've covered the essential details, there is a section on Alexa Calling and Messaging FAQs located here: www.amzn.to/2xy8nJ1

Kids Skills & Safety

On the **Things to Try** page you'll also see a link to **Kids Skills**; this provides an easy link up to all the Skills that have been designed with children in mind, such as story and joke telling, games and quizzes.

Access to Kids Skills are turned off by default, so to access them on your Echo you'll first need to **Enable Kids Skills**. Once enabled you can then **Explore Kids Skills**, and then enable each individual Skill you like to work with your Echo device.

Note that when you enable access to Kids Skills you are automatically allowing Amazon to store data from the Skill when it is used including voice-recordings. As with other voice interactions with Alexa, you can delete them by going to **Settings > History**.

Other things to consider concerning the safe use of your Echo when children are present, is to turn off **Voice Purchasing** in your shopping preferences and to turn on **Explicit Filter** in your music preferences! Both of these options can be found and managed in the Settings.

13: Alexa & Echo: Your Perfect Fit — Settings

For any voice command you read in this chapter,
be sure to use your wake word ("Alexa", "Amazon", etc) first.

The Settings section is where you control how the Alexa App and your Echo function to suit your style.

Personally, I prefer to deal with settings functions via the Alexa App on your PC or Mac, where I feel the interface is clearer. The rest of the information in this chapter assumes that you are looking at the App on your computer, however you can find all of these options on the App via your mobile phone or tablet too.

Let's walk through the Settings to get familiar with their contents and how you can tailor them to suit you. Select **Settings** on the App Menu to be taken to the **Settings Main Page** where we'll get started.

Device Settings

The first thing you'll see is a list of the Alexa enabled devices that you own and linked to your Alexa account. The word "Online" appears beneath any device when it is connected to a Wi-Fi network.

Start by selecting your Echo from the list, and you'll be taken to a page for basic setup, organization and information.

So these are the settings for your particular device:

Wireless and Connected Devices: Click on the options here to check which Wi-Fi network your devise is on; pair your device to other Bluetooth devices and remote controls if needed; and link to new Alexa Gadgets which are currently still in development – the idea being to link fun tech accessories with other features in the App (for example, linking an outside bell to chime when the Alexa timer expires.)

Do Not Disturb: This feature keeps Alexa quiet including preventing incoming Calls, Messages, Drop-ins or Notifications. At any time, say, "Do not disturb," Alexa will say, "I won't disturb you." Alternatively, you can toggle the Do Not Disturb on and off within the App here.

Scheduled Off: Do Not Disturb is programmable using this feature. Here's how:

- On the Alexa App, select the ***Do Not Disturb*** and toggle it on. Then tap or click on ***Scheduled*** to input the times you want the Do Not Disturb to start and finish.

- If you want to change or cancel the times listed, on a PC or Mac choose ***Edit*** at the bottom and on a tablet or mobile, just tap on the time you want to change

- Note that the Do Not Disturb can only be set for daily operation; and that Alexa will continue to alert you to alarms and timers even if Do Not Disturb is on.

Sounds: This is where you customize what sounds your Echo makes, when it makes them and how loud they are.

- **Alarm, Timer and Notification Volume:** Slide the bar left for lower and right for higher, and Echo will give a tone at the new volume for you to evaluate.

- **Audio:** The explanation of notifications is just around the corner. This allows you to determine whether a tone is given when a notification hits your Notification List. I keep Audio toggled Off unless I'm eagerly awaiting a Notification about a package being sent out for delivery or a message returned (See Calling & Messaging).

- **Custom Sounds/Alarm:** This section was covered earlier in Reminders & Alarms, but in brief, you can select from celebrity-delivered alarm messages and custom sounds. The Default alarm shows initially; if you change it, the new choice is shown in the box.

Device Name: Once you name your Echo devices on your account, the name will appear on the Settings home page and here:

- Select Edit.
- Change the Name.
- Select Save.

Device Location: I've included my complete address to get the best information about distances to restaurants and businesses near me. You don't have to enter any location, or you can input just a street and/or zip code:

- Select the Edit button.
- Input or change address information.
- Choose Save or Cancel.

Device Time Zone: To register the correct time on your Echo:

- Select a Region or United States from the top box to see your time zone options.
- Select the time zone you want from the lower box.

Wake Word: The default wake word is Alexa, but you can change it to Computer, Amazon or Echo. I've tried them all, and Echo is the only viable alternative for me since I often use Amazon and Computer in conversation unrelated to Alexa/Echo. Doing that creates a false "waking" of the device. If you want to change the wake word:

- In the Alexa App go to **Settings > Your Echo > Wake Word**
- Expand the Names box with the down caret.
- Select the name you want.
- Save it, and you'll be taken to the previous page where the Wake Word will appear.

Measurement Units: If you prefer metric units for temperature and distance, toggle the buttons to On.

Device is registered to: CAUTION! Once you set up your Echo, as explained in an earlier section, your name will appear here. The note of caution relates to the option to the right Reset to Factory Defaults. This will deregister your Echo and I'm aware of only two times to do this:

- If you sell or give away the device.
- If you cannot get it to function properly in the future, Amazon recommends deregistering and reregistering it as a last resort.

Deregistering your device can also be done online, for example if you sell your Echo but forget to deregister it first. To deregister any Echo device:

- Go to www.amazon.com/mycd
- Select Your Devices.
- Select the box to the left of the appropriate device.
- In the popup window, select Deregister your device.

Alexa Preferences

Once you've finished with the device settings above click back to the main Settings page. Below **Devices** you will find **Alexa Preferences** and **Alexa Account**.

Alexa Preferences is where you need to go to set up various default options for a number of features, these include:

Music & Media: This section lists the accounts you've linked to Alexa, along with the user ID for the account, and those available to use with Alexa that don't require an account. If you haven't opened an account for these services or haven't yet linked your existing account to the Alexa App, you can do so by selecting any of the music services from the **Music, Video & Books** tab on the App menu. Complete instructions are given in Chapter 5 (page 34). Select any of the Music Services to:

- Unlink the account if you don't use the service or Echo any longer.
- Manage your account Settings by logging into your account on the service's site.
- Manage service-specific Settings such as Enabling Custom Stations on iHeartRadio.

- Learn tips for using the music service.

Select the **Choose Default Music Service** option to choose a default music library and music station. For example, my libraries are Amazon Music and Spotify, and Amazon is my default. When I say, "Play Blake Shelton," Alexa plays music available on Amazon. If I want the music I have in my Spotify collection, I request it — "Play Blake Shelton from Spotify."

Flash Briefing: Most news services create short news summaries hourly or several times per day, in many cases to play on radio. Many of those summaries are available for playing on you Echo whenever you want to hear them. Most are Skills, and we've discussed Skills at length. Together, those summaries you choose make your Flash Briefing. To create yours:

- Select ***Get More Flash Briefing Content***.
- Browse the list, and select the ones you want.
- Enable the Skill.
- Those you choose will be listed here.
- Toggle on those you want to play.
- Choose ***Edit*** Order to rearrange the order they are played.
- Say, "Play my Flash Briefing," "What's in the news" or similar, and Echo will deliver it.
- Say, "Skip" or "Next," and Alexa will move on to the next service in the briefing.

 have 11 services in my briefing, but I often hear the top story from a service and then skip to the next or skip the service altogether.

Traffic: This feature was covered in ***Things to Try*** chapter; your Traffic feature gives you commute times. Input a "From" location, "To" location and stops in between, if desired. Edit them to change your route, and then ask, "What's my commute?"

Sports Update: This feature was covered in the ***Things to Try*** chapter (page 85), but here's a summary. The feature allows you to follow professional teams and leagues from North America and Europe plus the football and basketball teams for most major colleges in the US:

- Use the Search your Teams box to find teams you want to follow.
- Click on those teams from the search results.
- They will appear in your list.
- Remove them by selecting the X.
- Say, "Sports update" to hear latest scores and upcoming games for the teams in your list that are currently in season.

Traffic: Also covered in Things to Try, your Traffic feature gives you commute times. Input a "From" location, "To" location and stops in between, if desired. Edit them to change your route, and then ask, "What's my commute?"

Calendars: Alexa can link to calendars from Google, Microsoft and Apple iCloud. Link your calendars, and add items to them with basic requests such as, "Add appointment at 3pm Thursday to my calendar" and hundreds of similar things. See **Check and Manage your Calendar** in this guide's Things to Try chapter.

Lists: Alexa manages a Shopping List and To-do List for you in the App. You can also link your lists on Alexa to several 3rd party list apps like **AnyList** and **Todoist**:

- Create a free basic or paid premium account on one or more of those sites.
- Select ***Settings > Alexa Preferences > Lists***.
- Choose a service and select **Link Account** to connect the account to Alexa.
- Sign into your Amazon account and the list service account.
- Go to the separate app of the chosen service on your phone and tap "Connect with Amazon Echo".
- Sign in.

If you already have an account with one of those services, connecting it with Alexa might make sense. If you don't, it's redundant if you have the Alexa App on your phone.

Alexa Account

The penultimate section in the Settings page is the Alexa Account section. Here is where you can manage features such as which Notifications and allowing the Kids Skills to be linked to your device. Here is also where you can manage certain other useful features, including:

Voice Purchasing: The details on this topic are covered at length in the ***Things to Try*** chapter (page 88) under "Shop Amazon (For Prime Members)". In short, Amazon Prime members with a US address and means of payment stored on Amazon can:

- Enable 1-click shopping on your Amazon account.
- Enable Voice Purchasing here.
- Choose a 4-digit code and require its use with Voice Purchasing.
- Order and reorder Amazon Prime-qualified items with hands-free ease.

Household Profile: If you haven't already created an Amazon Household, clicking there will take you to Amazon so you can do so. When you create an Amazon Household, you gain four advantages:

- Members share Amazon Prime benefits when Prime Sharing is enabled.
- All digital content can be shared between adults in the household in what Amazon calls a Family Library.
- Select titles can be shared and unshared with children too.
- Payment instruments/methods can be shared with other adults in the Household.

A household can include up to two adults and four children. As we noted earlier, creating and managing an Amazon Household is done through your Amazon account. Once logged into Amazon.com:

1. Hover over **Accounts & Lists** near the top of the page.

2. Select "**Your Account**".

3. In the "Shopping programs and rentals" box, select **Amazon Households** to go to the Manage Your Household / Your Amazon Household Benefits page.

4. Enable Prime Sharing, if desired.

5. Select "Add an Adult," and follow the instructions including providing their login information, so they will have to give it to you or be there with you to type it in.

6. And/or select "Add a Child," Create and Save their Profile, and follow the link to Manage Your Content and Devices to manage the content they can access.

7. Adults and children can be removed just as easily, and you can "leave" a household too.

Leaving and removing can be done in this section of the App too. Select yourself, and choose Leave; select another Household member, and choose Remove. Complete information about households and how to manage yours can be found on this page: www.amazon.com/myh/manage

History: Select this Setting to view your entire voice-interaction history with Alexa. Choose any line to see complete details or to delete the item. To delete all voice recordings and their associated Cards:

- Go to www.amazon.com/mycd
- Select **Your Devices**.
- Select the box to the left of the appropriate device.
- In the popup window, select **Manage voice recordings**, and **Delete** or **Cancel**.

The Settings page also has some useful information concerning **Alexa Privacy**, in particular a link where you can **Manage Skills Permissions**.

14: Resources Galore — Help & Feedback

This section of the Alexa App provides a wealth of information, though we've already covered much of it and referred to it frequently. When selecting an item from Things to Try, for example, you are usually taken here for the explanation. There are things worth mentioning about this section that haven't been covered yet in this guide.

When you click or tap on *Help & Feedback*, at the top you will see the options *Alexa Features*, Alexa Devices and *Alexa Companion Devices*. Here's what each option will give you:

Alexa Features – General Alexa User Guide

As you will have guessed, *Alexa Devices* and *Alexa Companion Devices* options will give you information that is particular to your specific device (more about this later). However, clicking or tapping on *Alexa Features* will produce another page with further sections to explore about the Alexa App that are common to all Alexa-enabled devices, including these ones which are worth looking at in more detail:

Alexa Smart Home: This is where you'll get further answers for setting up, controlling and managing your smart home devices, including a useful section for troubleshooting the most common problems.

Alexa Entertainment: Lots of helpful tips to do with playing Music, Video and Audiobooks via Alexa, but sections of special interest here are:

- **Play Multi-Room Music on Echo Devices** is must-reading if you do have more than one Echo device. There's information about using devices with the same Wake Word, which devices don't require a Wake Word (Amazon Tap, e.g.) and things you can do (like share Music between devices) and things you can't do (like play the same music on more than one device).

- We discussed voice-control of Fire TV in the Video chapter, but the *Use Your Alexa Device to Control Your Fire TV* section in the App offers more detail and will be useful for Fire TV users for linking and controlling content.

Alexa Conversations: Here is your resource for Calling, Messaging, Drop-In capabilities. The information in this section and its links are comprehensive. You'll find step-by-step instructions to follow for setup and use, if what we discussed earlier isn't clear.

Help Around the House with Alexa: his is where you'll find answers to questions concerning most productivity tasks such as setting alarms, timers, shopping, making lists, getting your traffic, flash briefing and calendar updates and so on.

Alexa Devices - Echo User Guides

To get specific information about the features of the Echo, you will need to return to the *Help & Feedback* homepage first then go to *Alexa Devices* where you'll find help and tips to manage features specific Echo devices, including for example those that are specific to Echo Devices with a Screen, as managing accessibility features for users with vision or hearing impairments.

Alexa Companion Devices

This section of *Help & Feedback* is for setting up/troubleshooting companion devices such as the Echo Sub, the Echo Link Amp, the Alexa Voice Remote or using Alexa on non-Echo devices such as the Fire TV.

Contact Us

I've had no issues with Echo that I couldn't solve using the abundance of information found in the Alexa App and on Amazon. com. Because of the good fortune I've had with Alexa and Echo, I haven't emailed or called Customer Service. However, if you have problems and can't find answers, don't hesitate to contact Amazon. Feedback is another issue. I have used the Send Feedback option seven times to date to give Amazon my thoughts on how to improve

Alexa/Echo performance. I bring this up to say that Amazon has always been responsive, so if you contact the company with issues, you can expect that a customer service representative will be in touch.

Legal

If you have questions about how Amazon will use your information or what legal parameters there are for using Alexa/Echo, scanning the Legal & Compliance section will provide you with answer

15: Going to the Source — Amazon Pages to Be Familiar With

My goal is to provide you with clear, concise information from the perspective of someone that uses Echo and Alexa every day, insight I don't often find when reading pages on Amazon. That's why I wrote this book. Still, there are pages on the website that might help you get the most from your Echo.

Alexa Help: This is a home page with links to information on the broad categories of things Alexa can provide — Music & Entertainment, News & Information, Questions & Answers and everything else covered in this book. http://amzn.to/2k5tsrI

Echo Help: This page (https://amzn.to/2DpcNHy), and its many links, overlaps with the Alexa Help page, but the emphasis is just on the Echo.

Echo Geek Videos: This page (https://amzn.to/2IK1vzF) provides general overview videos for many key setup and use tasks.

Before You Go

So, there you have it! I trust by now you have your Echo up and running and have become familiar with many of its features. Please drop me a line at cjandersentech@gmail.com if you require any further clarification.

And finally, positive reviews on Amazon.com make a huge difference to the success of independent authors, such as myself. If you found this guide helpful, I would be very grateful if you took a moment to leave a comment. Thank you.

Printed in Great Britain
by Amazon